Why We Disagree about Human Nature

# Why We Disagree about Human Nature

EDITED BY
Elizabeth Hannon
and Tim Lewens

UNIVERSITY PRESS

Great Clarendon Street, Oxford, OX2 6DP,
United Kingdom

Oxford University Press is a department of the University of Oxford.
It furthers the University's objective of excellence in research, scholarship,
and education by publishing worldwide. Oxford is a registered trade mark of
Oxford University Press in the UK and in certain other countries

© the several contributors 2018

The moral rights of the authors have been asserted

First Edition published in 2018

All rights reserved. No part of this publication may be reproduced, stored in
a retrieval system, or transmitted, in any form or by any means, without the
prior permission in writing of Oxford University Press, or as expressly permitted
by law, by licence or under terms agreed with the appropriate reprographics
rights organization. Enquiries concerning reproduction outside the scope of the
above should be sent to the Rights Department, Oxford University Press, at the
address above

You must not circulate this work in any other form
and you must impose this same condition on any acquirer

Published in the United States of America by Oxford University Press
198 Madison Avenue, New York, NY 10016, United States of America

British Library Cataloguing in Publication Data
Data available

Library of Congress Control Number: 2017963189

ISBN 978-0-19-882365-0

Links to third party websites are provided by Oxford in good faith and
for information only. Oxford disclaims any responsibility for the materials
contained in any third party website referenced in this work.

*For our friends in the SCINAT team: Riana Betzler, Adrian Boutel, Andrew Buskell, Christopher Clarke, and Sam Murison*

# Contents

*Acknowledgements* ix
*List of Contributors* xi

Introduction: The Faces of Human Nature 1
*Tim Lewens*

1. Doubling Down on the Nomological Notion of Human Nature 18
   *Edouard Machery*

2. Trait Bin and Trait Cluster Accounts of Human Nature 40
   *Grant Ramsey*

3. A Developmental Systems Account of Human Nature 58
   *Karola Stotz and Paul Griffiths*

4. Human Nature, Natural Pedagogy, and Evolutionary Causal Essentialism 76
   *Cecilia Heyes*

5. Human Nature: A Process Perspective 92
   *John Dupré*

6. Sceptical Reflections on Human Nature 108
   *Kim Sterelny*

7. The Social Construction of Human Nature 127
   *Kevin N. Laland and Gillian R. Brown*

8. The Use and Non-Use of the Human Nature Concept by Evolutionary Biologists 145
   *Peter J. Richerson*

9. Human Ontogenies as Historical Processes: An Anthropological Perspective 170
   *Christina Toren*

10. Divide and Conquer: The Authority of Nature and Why We Disagree about Human Nature 186
    *Maria Kronfeldner*

*Index* 207

# Acknowledgements

Our first debt is to the dozen researchers who have contributed their work to this book. *Why We Disagree about Human Nature* began life at a conference of the same name held in Cambridge at CRASSH—the Centre for Research in Arts, Social Sciences and Humanities—in December 2015. Not every talk given at that meeting made it to the final printed book, and not every chapter collected here was originally presented at the conference. Even so, the conference promoted rich dialogue between the contributors as their chapters evolved. Those who were not able to make it to Cambridge took time to read the other draft papers submitted, to correspond with the other authors, and to shape their work in light of our earlier collective discussion. Readers will notice the frequent cross-referencing that occurs throughout this book, and we think the collection is far better for the open conversation that characterized its genesis.

We must also thank CRASSH itself for providing the usual polished and professional support in putting on our conference, and Samuel Murison—at the time part of our ERC-funded team—first for helping to organize the meeting, and then for significant editorial assistance. We owe a debt to Mike Hulme, whose book *Why We Disagree about Climate Change* (2009) gave the template for this collection's title. One of us (Lewens) would like to thank the other (Hannon) for making the whole process of co-editing a pleasure. And one of us (Hannon) would like to thank the other (Lewens) for their immense generosity and support. Finally, we must thank the two funders who have made this work possible: first the European Research Council (grant no. 284,123), who supported the conference and a team of researchers on a project entitled 'A Science of Human Nature?', and more recently the John Templeton Foundation, who have supported Lewens's research in the final stages of completing this volume.

# List of Contributors

GILLIAN R. BROWN is a Reader in the School of Psychology and Neuroscience at the University of St Andrews. Her research focuses on the development and evolution of sex differences in behaviour in human beings and other mammals. She has authored over sixty articles and one book, *Sense and Nonsense: Evolutionary Perspectives on Human Behaviour* (Oxford University Press, 2011).

JOHN DUPRÉ is Professor of Philosophy of Science at the University of Exeter, and Director of Egenis, the Centre for the Study of Life Sciences. His most recent book is *Processes of Life: Essays in the Philosophy of Biology* (Oxford University Press, 2012).

PAUL GRIFFITHS is Professorial Research Fellow in the Charles Parkins Research Centre, University of Sydney, and Visiting Professor in the Egenis Centre for the Study of Life Sciences, University of Exeter. His recent books include *Genetics and Philosophy: An Introduction* (Cambridge University Press, 2013) and *Cycles of Contingency: Developmental Systems and Evolution* (MIT Press, 2001).

ELIZABETH HANNON is Senior Fellow and Associate Director of the Forum for European Philosophy. She is also Assistant Editor of the *British Journal for the Philosophy of Science*.

CECILIA HEYES is Senior Research Fellow in Theoretical Life Sciences, and Professor of Psychology, at All Souls College, University of Oxford. Her book *Cognitive Gadgets: The Cultural Evolution of Thinking* will be published by Harvard University Press in 2018.

MARIA KRONFELDNER is Associate Professor at the Central European University. She has published widely in the philosophy of the life and social sciences, and has been awarded the Karl Popper Essay Prize of the *British Journal for the Philosophy of Science* and *The Philosophical Quarterly* International Essay Prize. Her book *Darwinian Creativity and Memetics* (2011) is published by Routledge.

KEVIN N. LALAND is Professor of Behavioural and Evolutionary Biology at the University of St. Andrews. He is the author of over 200 articles on the topics of social learning, cultural evolution, and niche construction. His books include *Sense and Nonsense: Evolutionary Perspectives on Human Behaviour* (Oxford University Press, 2011), *Niche Construction* (Princeton University Press, 2003), and *Darwin's Unfinished Symphony: How Culture Made the Human Mind* (Princeton University Press, 2017).

TIM LEWENS is Professor of Philosophy of Science at the University of Cambridge. His recent books include *The Meaning of Science* (Penguin), *Biological Foundations of*

*Bioethics* (Oxford University Press), and *Cultural Evolution: Conceptual Challenges* (Oxford University Press), all published in 2015.

EDOUARD MACHERY is Distinguished Professor in the Department of History and Philosophy of Science at the University of Pittsburgh, the Director of the Center for Philosophy of Science at the University of Pittsburgh, a member of the Center for the Neural Basis of Cognition (University of Pittsburgh-Carnegie Mellon University), and an Adjunct Research Professor in the Institute for Social Research at the University of Michigan. His recent work includes *Arguing about Human Nature* (Routledge, 2013), *Current Controversies in Experimental Philosophy* (Routledge, 2014), and *Philosophy within Its Proper Bounds* (Oxford University Press, 2017).

GRANT RAMSEY is a BOFZAP Research Professor at the Institute of Philosophy, KU Leuven, Belgium. His work centres on the philosophical problems at the foundation of evolutionary biology. He has published widely in this area, as well as in the philosophy of animal behaviour, human nature, and the moral emotions. He runs the Ramsey Lab (theramseylab.org), a highly collaborative research group focused on issues in the philosophy of the life sciences.

PETER J. RICHERSON is Distinguished Professor Emeritus at the University of California–Davis. His research focuses on the processes of cultural evolution. He has written, with Robert Boyd, *Culture and the Evolutionary Process* (University of Chicago Press, 1985) and *Not by Genes Alone: How Culture Transformed Human Evolution* (University of Chicago Press, 2005), an introduction to cultural evolution. He has co-edited a book *Cultural Evolution* (MIT Press, 2013) with Morten Christiansen and is an author of over 200 articles and book chapters.

KIM STERELNY is an Australian philosopher based at the ANU who has always worked in the borderlands between philosophy and the natural sciences, especially the life sciences. His books include *Thought in a Hostile World* (Blackwell, 2008) and *The Evolved Apprentice* (MIT Press, 2012).

KAROLA STOTZ is Senior Lecturer in the Department of Philosophy, Macquarie University. She publishes on philosophical issues in evolutionary, developmental, and molecular biology, psychobiology, and cognition, and her recent publications include *Genetics and Philosophy: An Introduction* (Cambridge University Press, 2013).

CHRISTINA TOREN is Professor of Social Anthropology, University of St. Andrews, where her research focuses on Fiji and the Pacific, and Melanesia, particularly in relation to sociality, kinship, and ideas of the person. Her books include *Mind, Materiality, and History: Explorations in Fijian Ethnography* (Routledge, 1999) and *Making Sense of Hierarchy: Cognition as Social Process in Fiji* (Athlone, 1990).

# Introduction
## The Faces of Human Nature

*Tim Lewens*

## Introduction

Humans today disagree, or so it seems, about human nature—about whether there is such a thing, and about how the idea of human nature might help or hinder our thoughts and actions. Some might be tempted to assume that the question of whether human nature is a scientifically respectable notion divides natural and social scientists along neat disciplinary boundaries, with biologists and psychologists as supporters, historians and anthropologists as detractors. After all, by giving his book *The Blank Slate* the subtitle *The Modern Denial of Human Nature*, the cognitive scientist Steven Pinker signals to his readers that a proper evolutionary perspective teaches us that we are not blank slates and consequently that human nature should not be denied (Pinker 2003). On the other side of the divide, the anthropologist Marshall Sahlins has argued for the 'modest' conclusion that, as he puts it, 'Western civilisation has been largely constructed on a mistaken idea of human nature'—an idea that, in Sahlins's view, even 'endangers our existence' (Sahlins 2008).

The truth is that debates over human nature are more complex than this easy disciplinary division suggests. While it is undoubtedly the case that many evolutionary theorists have enthusiastically embraced human nature, we must remember that large numbers of evolutionists have also rejected it. The philosopher David Hull's (1986) modern classic 'On Human Nature' set the tone for numerous later assaults on the concept from writers well informed about evolution. It is also important to recognize the nuanced views on human nature that come from the side of the social sciences. Sahlins, for example, does not claim 'that there is no such nature', but instead he asserts that 'its mode of existence and social efficacy depends on the culture concerned'. In other words, he is concerned to argue alongside Clifford Geertz (whom he cites approvingly) that 'there is no such thing as human nature independent of culture' (Geertz 1973: 49), not that there is no such thing as human nature *tout court*.

The contributors to this collection take very different stances with regard to the idea of human nature. They come from the fields of psychology, the philosophy of science, social and biological anthropology, evolutionary theory, and the study of animal cognition. Some of them are 'human nature' enthusiasts, some are sceptics, and some say that human nature is a concept with many faces, each of which plays a role in its own investigative niche. Some want to eliminate the notion altogether, some think it unproblematic, others want to retain it with reforming modifications. Some say that human nature is a target for investigation that the human sciences cannot do without, others argue that the term does far more harm than good. The diverse perspectives articulated in this book help to explain why we disagree about human nature, and what, if anything, might resolve that disagreement.

## The Anti-Essentialist Consensus

Amid the broad variety of views over human nature expressed over the past thirty years or so, a few points of consensus emerge. Charles Darwin learnt from his eight-year study of barnacles that variation was ubiquitous in all species: 'Not only does every external character vary greatly in most of the species,' he wrote, 'but the internal parts very often vary to a surprising degree' (Darwin 1854: 155). He went so far as to assert that it is 'hopeless' to find any part or organ 'absolutely invariable in form or structure' (p. 155). Variability in all parts of all species is a primary fact of nature, says Darwin, and this ubiquitous variation is the fuel that powers natural selection.

More or less all recent commentators have accepted this basic claim about variability, and most have moved on to argue that it rules out strongly essentialist conceptions of human nature, and of species natures more generally. If literally every trait within a species is prone to vary, then it cannot be the case that we will find any single trait whose possession is essential for membership of that species.

This fact of ubiquitous variability has consequences for how biologists understand species membership. Modern biologists are notorious for their failure to agree on the definition of what a species is. Even so, Darwin's insistence on the ubiquity of variation within species has led most modern biologists to understand species not as groups of organisms with internal features—genes, for example—in common, but instead as twigs on the tree of life. (Even here we have only a partial consensus, because not every taxonomist subscribes to this view. For example, the phenetic approach to taxonomy, according to which species membership is decided by a form of all-things-considered similarity, still holds sway in some circles (Lewens 2012a).)

For most taxonomists today, species are segments of the great genealogical nexus, with the result that what matters for species membership—that is, what it takes for an organism to be a member of a species such as *Canis familiaris* (the domestic dog) or *Homo sapiens*—is at least partly determined by genealogical relations. Just as membership of the British Royal Family is decided by who one's parents are, rather than by one's internal or external bodily features, so membership of a larger genealogical unit like a biological species is also settled by appeal to ancestry.

It is consistent with the rejection of very strong forms of essentialism—for example, with the rejection of the idea that what it is to be human is to possess a crucial 'human' gene, or a distinctively 'human' form of gait, intelligence, language, technological facility, or whatever—that we might instead understand what makes an organism a human in terms of the possession of some reasonably large proportion of a cluster of such intrinsic properties (cf. Devitt 2008, 2010; Lewens 2012b). But the genealogical image of species as twigs on the tree of life rules out even this weaker form of essentialism, at least if we read it as a claim about sufficient conditions for species membership. Possession of a large proportion of a cluster of intrinsic properties cannot be sufficient for being a human—even though one might still try to argue that it is necessary—because ancestry matters, too.

## The Taxonomic Case against Human Nature

It is primarily for these taxonomic reasons that biologically informed theorists like David Hull (1986) and Michael Ghiselin (1997) argued in the 1980s and 1990s that human nature was, as Ghiselin put it, a 'superstition'. For Ghiselin, human nature would have to pick out that set of intrinsic properties that make us human. Since Ghiselin, like Hull, had argued that biological taxonomy didn't work that way, he thought that the study of evolution exposed the very idea of human nature as an anachronism. But evolutionists have not been united in their rejection of human nature. In 1990, for example, the evolutionary psychologists John Tooby and Leda Cosmides announced their intention to defend 'the concept of a universal human nature' (Tooby and Cosmides 1990), and we have already seen that Steven Pinker, too, holds that the deniers of human nature are misguided. These authors' adherence to the human nature concept, in spite of their endorsement of dominant taxonomic practice, indicates that the likes of Ghiselin and Hull may have been too demanding in their views of what job the notion of human nature is meant to do.

Biologists these days rarely feel the need to formulate a catalogue of properties that are found in all and only humans. But we can give up on that anachronistic taxonomic project while keeping alive the thought that some properties—psychological, physiological, and anatomical—are present in almost all humans, and that evolutionary processes explain why those properties became and remain so prevalent in our species. Perhaps a claim like 'Trichromatic vision is an element of human nature' commits us to nothing more than an assertion of widespread presence and its evolutionary rationale.

It is hardly news that humans differ in terms of how they see colours. We have known about colour blindness for hundreds of years. Even so, the great majority of humans have three classes of cone cell in their retinas, and are free of colour-blindness because of this. What is more, there is a plausible evolutionary story to be told about why this trichromatic form of vision emerged: John Mollon has argued that it was favoured by natural selection because it enabled our primate ancestors to detect ripe fruits against a dappled, leafy background (Mollon 1989). To say that trichromatic vision is part of human nature does not, on this view, imply that literally all humans are trichromatic, or that colour-blind people are not members of *Homo sapiens*.

We have seen a background consensus that lies behind a number of disagreements. Few modern thinkers have tried to defend what we might think of as taxonomic species natures—natures the possession of which make it the case that an individual is a member of one species rather than another. Even fewer have tried to defend strict essentialist pictures of species natures, understood as properties that are necessary and sufficient for species membership. But perhaps those evolutionary psychologists we quoted above, in setting themselves the target of describing human nature, are doing nothing more than listing the traits that most humans have, by virtue of the evolution of our species (see Machery 2008). What could be wrong with that?

## The Field-Guide Conception

23 January 2003 turned out to be the last day that NASA would receive a signal from the Pioneer 10 probe (NASA 2007). Even then, the spacecraft was already well beyond the limits of the planetary solar system, at an estimated distance of 12 billion kilometres from the Earth. Pioneer 10 had been launched in March 1972, and in anticipation of the possibility that it might eventually travel far enough to fall into the hands of intelligent aliens, a six-by-nine-inch gold-anodized aluminium plaque had been bolted to the probe's frame (Macauley 2010).

The plaque, which is reproduced on this book's cover, featured various images and diagrams intended to indicate the existence, intelligence, and location of the senders to alien receivers. There was also a picture of a human woman and man. This last element of the Pioneer 10 plaque gives us a useful jumping-off point to understand some of the reasons why—in spite of a broad anti-essentialist consensus—there remains vigorous disagreement about the very idea of human nature.

Some might argue that there is no way to attack the modest notion of human nature sometimes favoured by evolutionary psychologists without also denying that humans are part of the more general evolutionary order of things. Consider that we produce stylized representations of the members of all sorts of non-human species in field guides. My copy of the *Collins Bird Guide* (Svensson et al. 2010) features what are, to all appearances, entirely unobjectionable images of a typical male blackbird alongside a typical female blackbird. So why shouldn't we send aliens images of the typical human male and female? Carl Sagan and Frank Drake had designed the other elements of the Pioneer 10 plaque, and—partly because of her training in fine art, partly because of her availability at short notice—they asked Linda Salzman Sagan (Carl's wife) to produce the drawings of the humans (Macauley 2010: 108). She aimed to make her drawings 'representative of us all' (Salzman Sagan, quoted in Macauley 2010: 109).

The kind of generic images that feature in field guides give readers the ability to tell blackbirds from starlings; and, on the face of things, nothing stands in the way of including the species *Homo sapiens* in a field guide, enabling alien visitors to correctly pick out humans among the bewildering variety of organisms they will encounter on first visiting our planet. It is hard to deny that field guides do their job well, that the generic images they feature are valuable for assigning individuals to the correct species, and that they can do this in spite of the equally hard-to-deny fact that every individual organism—whether human, starling, or blackbird—is strictly unique.

Quite a few philosophers have been tempted by this 'field-guide' conception of human nature. Edouard Machery once wrote that 'describing human nature is [...] equivalent to what ornithologists do when they characterize the typical properties of birds in bird fieldguides' (2008: 323). More recently, Peter Godfrey-Smith also notes that Martian visitors would surely want an entry on humans in their field guide to Earth, and adds that 'In that sense, there is surely nothing mythical about the idea of human nature' (2013: 140). We saw a little earlier that Ghiselin's argument against human nature turned in part on the variability, in all respects, that characterizes the members of every species. The appeal to field guides reminds us that images of general attributes of species members can have diagnostic power in spite of this pervasive variability.

The Pioneer 10 plaque helps us to understand both the strengths and the limitations of the field-guide conception of human nature. It is indeed uncontroversial that we can supply curious observers with diagnostic images that enable them to sort individuals into species with considerable, albeit imperfect, success. But this falls short of a full-blown specification of the 'nature' of a species. After all, diagnosis relies on using features that are relatively easy to observe. Field guides tend to describe where a bird is most likely to be seen, its plumage, its song, and so forth. They do not describe aspects of a bird's behaviour that are unlikely ever to be seen by observers, and they do not describe internal features of anatomy or physiology, no matter how typical they might be for the species in question. Importantly, field guides often attempt to describe the typical song of a given species, regardless of whether ornithologists consider that song to be innate or learned.

There is a long tradition of exploring the varying abilities of birds to learn both local song 'dialects' and the songs that are typical of alien bird species (e.g. Thorpe 1958). So while a field guide *does* usually aim at giving us a description of a species' nature understood as that which we can use to tell one species from another, it does *not* aim at giving a description of a species' nature understood as that which is independent of what we might think of as the species' 'culture'. And so we can reasonably ask why, if the Pioneer 10 plaque's images of a man and a woman are simply supposed to tell a travelling alien what to look out for on a visit, are the man and woman both shown naked? For the great majority of people across the world, the norm is to spend most of their time (when they are observable to casual onlookers) wearing clothes. Visiting groups of aliens will arrive with a mistaken impression of our usual appearance if they use the Pioneer 10 plaque as a basis for their expectations. Showing us without clothes is akin to refusing to describe bird song in field guides, in those cases where we worry that the song might be learnt.

## Three Roles for Human Nature

Our discussion highlights that we might want to use the notion of human nature to do at least three jobs. We have already seen that human nature has a diagnostic function for some thinkers: the 'nature' of a species is something that observers can make use of in assigning organisms to the species of which they are members. This is the view of human nature that underlies the field-guide conception. A second function is also comparative: in asking whether humans are, in general, similar or dissimilar to the members of other species, we can also talk in terms of human nature compared with the nature of chimps, or dogs, or mice. In other words, in talking of human nature we simply talk of what humans are typically like. The third function requires that we consider human nature in a manner that contrasts it with 'human culture'. It asks us not merely to generalize about what humans are like, but to divide human traits into two categories. This way of thinking of human nature invites us to consider whether a widespread human trait—it could be anything from the ability to imitate the bodily movements of others to the susceptibility to visual illusions—is truly part of our nature, or whether we might instead attribute it to some form of cultural influence.

The second species-comparative view differs from the diagnostic view. First, aspects of human cognition that are very hard for casual observers to detect, such as our tendencies to make mistakes in probabilistic reasoning, become candidates for human nature so long as they are indeed typical of humans, even though their hiddenness makes them unlikely to appear in a hypothetical field guide. Second, aspects of human cognition or behaviour that also appear regularly in animals—perhaps the broad syndrome of neural and physiological reactions that underlie fear responses—also become candidate features of human nature, even if (as Darwin thought for fear) they are also features of the nature of dogs. Traits that are common to humans can, on this view, be elements of human nature in spite of the fact that their not being unique to humans makes them useless for the purposes of species diagnosis.

It is telling that Cosmides and Tooby characterize their project—which, as we have seen, they equate with providing a description of human nature—as the equivalent of providing a version of *Gray's Anatomy* that deals with human psychology (Tooby and Cosmides 1992: 68–9). Like a field guide, the original *Gray's Anatomy* was dedicated to the provision of practical knowledge. But Henry Gray's intended purpose for that knowledge was of a wholly different kind:

> This work is intended to furnish the Student and Practitioner with an accurate view of the Anatomy of the Human Body, and more especially the application of this science to Practical Surgery. (Gray 1858: vii)

Hence *Gray's Anatomy*, unlike a field guide, describes hidden structures, and also structures that are very similar in humans compared with other creatures. Its purpose was not to allow species recognition, but to present a generic form of knowledge about typical humans in a way that would enable medical students—especially surgeons—to learn and practise efficiently. By analogy, a version of human nature that might be articulated in a '*Gray's Anatomy* of the mind' would simply tell us what typical humans are like with respect to their cognitive traits.

It is important to keep this second notion of human nature in mind, not only because it seems relatively uncontroversial to say that there are some fairly general truths about what more or less all humans are like—we are prone to predictable fallacies in probabilistic reasoning, we cannot see light in the ultraviolet range, we sweat when fearful, we frequently act for the sake of unrelated others—but also because there is reason to think that a notion very much like this one is at work in such well-known works as David Hume's *Treatise on Human Nature*.

The historian of philosophy Peter Kail (Kail 2012) has drawn our attention to an essay by Hume of 1741, entitled 'Of the Dignity or Meanness of Human Nature', in which Hume notes that in 'forming our notion of human nature, we are apt to make a comparison between men and animals, the only creatures endowed with thought that fall under our senses' (Hume 1995: 82). But while these forms of comparison help us to delineate human nature, human nature is *not* understood as that which makes us different from animals. This is not a diagnostic notion of human nature, and neither is it a notion according to which human nature names that which is *distinctively* human. Indeed, Hume's view—later expressed in the *Treatise on Human Nature*—is that commonalities in behavioural 'comportment' between humans and animals ultimately suggest to him that the mechanisms that underlie perception, cognition, and action are the same in our own species and in animals: in the 'whole sensitive creation [...] every thing is conducted by springs and principles, which are not peculiar to man, or any one species of animals' (Hume 1978: 397). It is not contradictory, on this conception, to say that human nature—what humans are typically like—is fundamentally the same as animal nature in many respects.

This brings us to the third role for human nature. Reflection on Hume reminds us of the differences between using 'human nature' merely to mark out a series of traits that are typical for our species, and using 'human nature' as a foil for 'human culture'. In Hume's case, there is evidence that he does use 'human nature' in the former sense, while refusing to use it in the latter sense. For example, Hume suggests that impersonal norms of justice are established by convention. Roughly speaking, his view is that humans collectively institute conventions such as respect for private property because of the advantages these conventions bring to animals like us that live in social groups. Even so, education teaches respect for these conventions, and this leads to an altered set of underlying motivational dispositions. Kail highlights Hume's assertion that feelings of justice take on such 'firmness and solidity, that they may fall little short of those principles which are the most essential to our natures, and the most deeply radicated in our internal constitution' (Hume 1978: 501). In other words, 'human nature' names that which is 'firm and solid', or that which has deep roots in our 'internal constitution', regardless of what processes—social convention, education, or otherwise—explain such traits' entrenchment.

## The Perils of Human Nature

Human nature is a notion that is ripe for abuse. It can be used to entrench—as part of what Hume described as the 'deeply radicated' order of things, 'firm and solid' and hard to shift—that which is in fact quite amenable to change. The Pioneer 10 images that purport—in Salzman Sagan's words—to be 'representative of us all' show the male figure with his hand raised in greeting, apparently taking the lead in approaching the oncoming alien. The female figure instead keeps her arms by her side, and seems turned slightly towards the man, giving the impression that while the man takes the job of addressing the alien, the woman looks at the man. It is not surprising that this effort to represent the general state of our species drew ire from some feminists at the time.

NASA's website tells us how these supposedly pan-human images were generated: 'The physical makeup of the man and woman were determined from results of a computerized analysis of the average person in our civilization' (NASA 2007). In sharp contrast with this official line, Salzman Sagan's own recollections of how she made the images suggest something more low-tech:

And I took some of my art books—I didn't have anybody pose or anything—they weren't done from life—but I looked at Greek statuary because that was what was readily available in terms of being more realistic than, for example, a figure from Van Gogh or Gauguin or anything. So, I looked at Greek statuary. (Salzman Sagan, quoted in Macauley 2010: 112)

Salzman Sagan began with a culturally specific, idealized image of a beautiful human body. But she then added to the ideal form of Greek statuary the stereotypical features of various 'races' in a mix-and-match approach:

I did try to make them interracial. So if the man had curly hair the woman had straight hair. There was a slant to the woman's eyes—it was fun because it was the kind of thing where you want to show as much interracial features as you could and kind of make them representative of all of us. (Macauley 2010: 109)

Carl Sagan's recollections suggested that these intentions were not quite seen through in the final versions. There was an initial effort to 'have the man and woman panracial', including such things as an 'Afro' haircut for the man and an 'Asian' appearance for the woman. Unfortunately:

[…] because the woman's hair is drawn only in outline, it appears to many viewers as blond, thereby destroying the possibility of a significant contribution from an Asian gene pool. Also, somewhere in the transcription from the original sketch drawing to the final engraving the Afro was transmuted into a very non-African Mediterranean-curly haircut. (Sagan 1973: 26)

Sagan nonetheless felt confident that 'the man and woman on the plaque are, to a significant degree, representative of the sexes and races of mankind' (Sagan 1973: 26–7).

Since humans from around the world are a diverse lot, there is also a risk that we inadvertently take that which is, in fact, typical only of a subsection of the human population that we happen to be familiar with, and claim that it is a part of human nature in the sense that it holds good for the species as a whole. Anthropologists have long been concerned about basing claims about psychological universals on the responses of students from Western universities. These worries have received plenty of attention since a publication by Henrich et al. (2010) pointing out the misleading effects of psychologists' reliance on what they call WEIRD (Western, Educated, Industrialized, Rich, Democratic) research subjects. Henrich and colleagues give good empirical reasons to worry that what psychologists might find typical of economic reasoning, moral intuitions, even perceptual capacities in this subset of persons is not typical of people in general.

Henrich et al. have indeed done an important service in alerting cognitive scientists to these forms of diversity, but it is worth pointing out that similar worries have been raised within the community of social anthropologists for decades. Philippe Descola, for example, remarked as early as 2005 (the work was not translated into English until 2013) that supposed cognitive universals have often been inferred 'on the basis of experiments conducted almost exclusively in Western industrialized societies' (2013: 102). And way back in 1901, the anthropologist W. H. R. Rivers produced evidence suggesting that susceptibility to visual illusions depended on whether you were from Cambridge (as his students were) or the islands of the Torres Straits on the other side of the world (Rivers 1901). These worries do not threaten the very idea of human nature, but they do remind us of the hurdles that need to be cleared before we can confidently make claims about what is typical of humanity in general.

## What Lies Ahead

The disputes about human nature briefly addressed in the preceding pages play out across the contributions to this volume. The first four chapters all offer defences of the notion of human nature, but each does so in a markedly different way. Later chapters either express outright scepticism about the value of the idea of human nature, or they offer explanations for why that notion has been—and continues to be—contested.

### Machery

Edouard Machery's opening chapter defends his important 'nomological' account of human nature against recent criticisms. In its original (2008) formulation, that account proposed that 'Human nature is the set of properties that humans tend to possess as a result of the evolution of their species'. Machery, consistent with much earlier suggestions from the likes of David Hume, proposes that for a trait to be a part of human nature it needn't be literally universal—that is, present in every human—but it must be at least typical of humans. Machery adds that not just any trait that happens to be widely held among members of *Homo sapiens* at some moment in time is a part of human nature. If a trait owes its commonality to what Machery classifies as a non-evolutionary process—he considers the example of songs that go 'viral' to the extent that they become known by more or less everyone on the planet—then it is not part of human nature.

Machery's original account invited two broad kinds of attack. First, what is the rationale for insisting that only more or less universal traits can be part of human nature? This seems especially pressing in the context of evolutionary sciences, where the very same processes—natural selection, mutation, migration, genetic drift—are invoked to explain why traits are present at low, middling, and high levels in populations, and why there can be stable mixtures of alternative traits. From an evolutionary perspective—or, at the very least, from the perspective of population genetics—it seems that there is nothing of theoretical significance about near-universality.

Machery's answer to this problem is to point out that the rationale for linking human nature to that which is typical may derive from considerations that lie outside the theoretical preoccupations of evolutionary biology. Even if there is no special *evolutionary* significance to be found in traits that are present at close to 100 per cent, compared with traits that are maintained by selection at much lower frequencies, evolutionary theory does not deny that some traits are, as a matter of fact, very widely shared among all humans. If we look away from evolutionary work, it seems that disciplines like physiology and medicine have good reason to focus their attentions on typical minds and bodies.

Machery's comments here are bolstered by our earlier discussion of *Gray's Anatomy*, and Cosmides and Tooby's conviction that in describing human nature we offer a *Gray's Anatomy* of the mind. Medical practitioners need to understand how human bodies work if they are to intervene in them effectively. Since the cognitive capacities of medical students (as well as the investigative capacities of anatomists and physiologists) are finite, it makes sense to prioritize anatomical and physiological education that focuses on those traits that one can reasonably expect to be present in any patient the practitioner might encounter. Such education is then augmented with further descriptions of broad subdivisions of our species, when *Gray's Anatomy* moves on to describe that which is anatomically typical for males and females.

More recent editions of *Gray's* have been further swollen by notes on common anatomical variations, and students are sometimes referred to the online version of *Gray's* for fuller details of these complicating matters. On this view, a description of human nature, understood as a version of *Gray's Anatomy* that deals with cognition, would constitute a comparatively easy-to-learn toolkit enabling reasonably successful interaction with any human one might happen to be presented with. As with the original *Gray's Anatomy*, that project leaves room for additional documentation that would capture the full detail of human cognitive variability.

Machery also responds to the second main question prompted by his 2008 proposal: how are we to say which processes are 'evolutionary'? Several theorists—represented in this collection by Peter Richerson, Kevin Laland, Gillian Brown, Cecilia Heyes, Kim Sterelny, and others—have argued that learning from others has been an important process in enabling us to adapt to our surroundings. They therefore wish to think of cultural processes as properly 'evolutionary'.

Machery responds with a mixture of pluralism and pragmatism. He suggests that there is indeed a perfectly reasonable sense in which we can count social learning as an 'evolutionary' process. However, he maintains that there are good reasons for understanding human nature in a narrower way. We should restrict our understanding of evolutionary processes in this context to those that act on genetic variation, and thereby exclude cultural processes, precisely because this maintains a less baggy, more focused account of what human nature is. Machery's thought is that it would undermine the practical value of the concept of human nature if—because we had become convinced that social learning was itself an evolutionary process—we were forced to conclude that when it becomes typical of humans to recognize the face of David Beckham, this ability becomes part of human nature.

*Ramsey*

Grant Ramsey offers a defence of his own 'trait cluster' account of human nature, and in so doing he argues for advantages of his own position over Machery's. Ramsey thinks of Machery as offering a 'trait bin' account. The name of this game is to provide a criterion that allows us to sort human traits into those that belong to human nature and those that belong elsewhere. The 'elsewhere' may be culture, or nurture, or pathology, or some alternative 'bin'. Ramsey suggests that this starting point is mistaken. We should not attempt to construct a divisive notion of human nature that rules some traits in and others out. Instead, we should understand 'human nature' as a name for the patterns that can be seen in potential developmental trajectories across human lives.

For every developing individual, there are some life courses that are comparatively likely, others comparatively unlikely, and others impossible. So each individual has a 'nature', in the sense of a pattern of outcomes open to that person. Schooling is likely to be under way by the age of 8 for most individuals, greying of the hair usually comes after one's 20s, and death comes before the age of 110. Evidently the patterns of possibility differ with respect to all such outcomes depending on where in the world one is born, and what one's social and genetic endowment might be. But if we add up the patterns that constitute individual natures, then we arrive, says Ramsey, at human nature. Human nature, for Ramsey, simply is that general set of patterns across all individual human potential life courses. Moreover—as the example of schooling indicates—Ramsey has no intention to sort human nature from human culture. The patterns that make up human nature can be discerned in traits of all kinds.

*Stotz and Griffiths*

Karola Stotz and Paul Griffiths offer a third account of human nature that, as they note, has significant commonalities with Ramsey's. Like Ramsey—and unlike Machery—they are not in the business of offering an account of human nature that is meant to be opposed to human culture. And, like Ramsey's, their account of human nature focuses on human variability and human plasticity.

For Stotz and Griffiths, the key to formulating a defensible account of human nature is to look to developmental processes. Since these processes can involve learning and cultural tradition, and since these processes are inherently variable with respect both to their constituent sub-processes and to their overall phenotypic outcomes, the account of human nature they yield neither stands in contrast to human culture nor emphasizes that which is common to all humans. Even so, Stotz and Griffiths argue that this counts as a full-blooded account of what human nature is, not least because in equating human nature with underlying human developmental processes, we arrive at an explanatory conception of human nature. Human nature, for Stotz and Griffiths, is that which explains more superficial elements of human behaviour. Moreover, their account preserves the thought that human nature is that which the human sciences—and here they include everything from social anthropology and economics to psychology and physiology—aim to delineate.

## Heyes

Cecilia Heyes offers the collection's fourth and final positive proposal for how to understand human nature, which she calls 'evolutionary causal essentialism'. In common with Machery, Heyes proposes a 'trait bin' account: her aim is to give a criterion that counts some traits as elements of human nature, while ruling others out. And, like Machery, Heyes argues that for a trait to be part of human nature, (i) it must be more or less universal within our species, and (ii) the trait must owe its presence to an evolutionary process. In contrast to Machery, Heyes shares with Stotz and Griffiths the requirement that human nature must be explanatory of more superficial behaviour. She therefore equates human nature not with any old trait that happens to be widespread within *Homo sapiens*, but instead with the 'set of mechanisms that underlie the manifestation of species-typical cognitive and behavioural regularities'. In this respect (and here she is influenced by earlier work by Richard Samuels (2012)), her account is more restrictive than Machery's.

In another important respect Heyes's account is more liberal than Machery's, because she embraces a broader view of what evolutionary processes are. She counts among 'evolutionary' processes those that are mediated by cultural and epigenetic inheritance. This means that while Heyes's account does indeed propose two 'bins' into which we can sort traits, we will not find one labelled 'nature' and the other 'culture'. Her view is that some of the traits that are part of human nature are present because of cultural inheritance and cultural evolution.

On Heyes's view, traits that fall into the 'non-human nature' bin include those merely behavioural traits that are not 'deep' enough to constitute explanatory mechanisms, those that are not widespread enough to be typical of the species, and those that do not have the right kind of history to count as products of evolution. Heyes builds to this conclusion via a critical appraisal of Gergely and Csibra's influential theory of 'natural pedagogy', according to which human infants are equipped with a distinctive set of adaptations that make them receptive to teaching (e.g. Csibra and Gergely 2011). Heyes's view is that the mechanisms that underlie our abilities to learn via teaching are themselves the products of cultural inheritance; moreover, the very fact that such distinctive, important, and pervasive human abilities owe their presence to a form of cultural evolution helps bolster her case for thinking that human nature should be defined by reference to a liberal collection of evolutionary processes.

*Dupré*

John Dupré's contribution is the first in the collection to express a more sceptical stance on human nature. His chapter briefly summarizes his process-oriented view of life itself, before applying that perspective to the specific question of human nature. An organism, says Dupré, is not best understood as a static object; rather, it is best understood as a process, or a collection of processes. An organism is something that changes over its life-course, and the processes of change that it undergoes are themselves the upshots of numerous constituent sub-processes. Like Ramsey, and Stotz and Griffiths, Dupré also stresses the plasticity of all of these processes. He suggests that by virtue of being sequences of events rather than persisting things, these life processes are liable to reach multiple alternative end-points via multiple intermediary stages.

Dupré worries that thinking in terms of a shared human nature encourages a tendency to focus on a single adult developmental stage as representing the 'nature' of our species. His own processual image of organisms instead downplays the idea of any single developmental stage as constitutive of what a human is, while also drawing attention to the flexibility of human developmental pathways. Dupré's overall stance is thus balanced between a positive proposal to understand human nature in terms of flexible developmental processes and a more sceptical position that rejects talk of human nature altogether.

*Sterelny*

Kim Sterelny's chapter offers a pragmatic rationale for scepticism of human nature. He agrees with some of the defenders of human nature that there may well be a reasonably rich set of traits that a large majority of humans share, and which are distinctive of the human species. (His suggested candidates include various abilities related to language use, moralizing, theoretical reasoning, and empathizing.) Even so, asks Sterelny, what would such a conception of human nature be good for? What valuable role would it play?

Sterelny gives several reasons for thinking that there may be no useful role for human nature. An account that focuses on commonalities held across all humans is of little value from an evolutionary perspective, because evolutionary thinking teaches us to focus as much on variation as on homogeneity. An account that focuses on what we owe to genetic evolution is challenged by the coevolution in our own lineage of genetic and cultural forms of inheritance. Sterelny's concerns extend as far as the constraining role sometimes accorded to human nature. The idea he targets is that if we know something about realistic limits of individual human cognitive, social, and behavioural capacities—if, that is, we know about human nature—then we will understand what sorts of general political systems and social arrangements are feasible, what sorts of specific social policies are likely to succeed in their aims, and so forth.

In response, Sterelny cautions that the notion of human nature can obscure the reciprocal manner in which alternative social arrangements can constrain the development of individual psychologies just as much as individual psychologies constrain social arrangements. He also argues that the question of which forms of political and social organization are feasible or harmonious is often explained not by generic facts about the typical psychology of individuals, but by facts about how societies instantiate mixtures of different psychological profiles. So if we want to understand what forms of social organization are achievable without miraculous intervention, we will not be assisted by assuming that human nature corresponds to that which is psychologically more-or-less universal.

### Laland and Brown

Like Sterelny, Kevin Laland and Gillian Brown are human-nature sceptics. And like Sterelny, their scepticism is pragmatic: they see no useful role for the human nature concept to play, and plenty of misleading ones. Their view is that human nature is a social construct, and in two senses. The first is comparatively obvious: the *concept* of human nature is, of course, a human invention, as are all concepts. But they also argue that human nature itself—if we mean by that the pattern of robust tendencies shown again and again across the individual members of our species—is quite literally constructed by social relations. They draw on a variety of processes stressed by those who seek 'extensions' to the largely gene-centric image of evolution associated with the Modern Synthesis—processes like niche construction, non-genetic inheritance, gene–culture coevolution, and so forth—to argue that there is no biological sense in trying to separate that which we owe to 'nature' from that which we owe to 'culture' or the environment. Since, in their view, the past activities of earlier generations of organisms always play some role in structuring the developmental environments of plants and animals, it simply makes no sense to think of some traits as free from forms of social influence.

We have already seen in the remarks at the beginning of this introductory essay that this role of demarcation is not the only one we might assign to human nature. But Laland and Brown worry that even if we try to reserve the term for other roles, it is nonetheless liable to be badly misunderstood. Their final judgement is that 'human nature' is a term we would be better off without.

### Richerson

Peter Richerson's essay also has a pragmatic focus, and he again finishes with a largely sceptical view of human nature. He aligns himself with Laland and Brown's conviction that the Modern Synthesis in evolution needs to be extended to include a more varied set of adaptive processes than natural selection acting on genetic variation. But he approaches our topic in a more historical manner, by looking at the ways in which a series of the most prominent evolutionary thinkers from Darwin to contemporary researchers have made use of the notion of human nature.

Richerson is not a blanket sceptic: he has no qualms about the basic, comparative use of 'human nature'—akin to the Humean notion outlined earlier in this introduction— that enables us to make general claims about what members of *Homo sapiens* are typically like, and the ways in which they may be similar to, and different from, the typical members of other species. Instead, Richerson draws on his own extensive body of research in cultural evolution to stress not only that various forms of social learning constitute transformative, creative, and adaptive processes currently acting on humans, but that these same learning processes have affected our ancestors over long swathes of evolutionary history. Richerson suggests not merely that human nature is a moving target— something that has been subject to change over time at the hands of cultural forces—but that human nature has always been constituted by mutually interacting cultural and genetic factors. His scepticism of the human-nature concept derives from what he sees as a series of misguided efforts—not present in the work of Darwin, but increasingly prevalent through Modern Synthesis versions of evolutionary theory—to distinguish between a pre-cultural human nature and a culturally overlaid 'superorganic' realm.

*Toren*

Christina Toren's manifesto for a thoroughly historical perspective on change and stasis in human individuals and groups has much in common both with the processual stance on human nature advocated by John Dupré and with the constructivist account offered by Laland and Brown. Toren, like many of the contributors, aims to do away with the notion that we can think of some traits as products of nature, rather than of culture. Toren argues—much like Laland and Brown—that developmental processes invariably proceed as they do by virtue of the structured array of developmental resources bequeathed by the interactions of earlier generations. For Toren, too, the developmental niche is a social product.

There are, however, significant differences between Toren's account and that of some of our other human-nature sceptics. If it makes no sense to single out human nature in contrast with human culture, then equally it makes no sense to point to human culture as something to be contrasted with human nature (see Lewens 2017). Rather than drawing on the importance of distinctively *cultural* evolutionary processes to undermine the idea of human nature, Toren consequently seeks a framework for understanding human evolution at all scales that uses neither the notion of 'nature' nor the notion of 'culture'. It is at this point that her affinities with John Dupré become clear: Toren argues that we should understand evolution as nothing more than the sum of microhistorical ontogenetic processes, in which the nature/culture divide disappears altogether.

## Kronfeldner

While most of our contributors—with their diverse proposals for how, if at all, we should understand human nature—*exemplify* disagreement, the final contribution from Maria Kronfeldner fittingly closes this collection by trying to *explain* this disagreement. Instead of offering a single account of what human nature is, Kronfeldner instead argues that the term has a number of quite different epistemic roles. She argues that since 'human nature' has so often been used to make assertions regarding that which is important regarding our own species, it is hardly surprising that the notion is continually subject to disagreement. Moreover, 'human nature' has often been used in a related way to imply that some individuals who lack that nature are less than fully human. Again, disagreement over what human nature is should hardly be surprising, given Kronfeldner's suggestion that the term itself is used to exclude or marginalize. Kronfeldner does not claim that modern scientific researchers continue to use 'human nature' in these exclusionary ways, but this does not undermine her conviction that we should expect disagreement to continue regarding what human nature is.

The earlier chapters in this volume point repeatedly to the question of what pragmatic role 'human nature' is supposed to play. Kronfeldner seizes on this to argue that since there are numerous different and defensible functions that scientists may wish to assign to the term, we should expect disagreement in how the term is used to persist. To give just a few examples, scientists may be aiming at a characterization of recent humans, of *Homo sapiens* in its entirety, or of broader segments of the hominin lineage. They may aim at describing robust psychological tendencies of typical individuals, or perhaps at describing social networks and the ways in which diverse individuals are required to sustain those networks. Depending on whether scientists are interested in features of social life that characterize very recent history or longer swathes of time, they may reckon different social and psychological features 'important'. If Kronfeldner is right, then disagreement about human nature will persist, even after readers have fully digested the contents of this book.

## References

Csibra, G., and Gergely, G. (2011). 'Natural Pedagogy as Evolutionary Adaptation.' *Philosophical Transactions of the Royal Society B* 366: 1149–57.
Darwin, C. (1854). *A Monograph on the Sub-Class Cirripedia*, vol. 2. London: Ray Society.
Descola, P. (2013). *Beyond Nature and Culture*. Chicago: University of Chicago Press.
Devitt, M. (2008). 'Resurrecting Biological Essentialism.' *Philosophy of Science* 75: 344–82.
Devitt, M. (2010). 'Species Have (Partly) Intrinsic Essences.' *Philosophy of Science* 77: 648–61.
Geertz, C. (1973). *The Interpretation of Cultures*. New York: Basic Books.
Ghiselin, M. T. (1997). *Metaphysics and the Origins of Species*. Albany: SUNY Press.
Godfrey-Smith, P. (2014). *Philosophy of Biology*. Princeton, NY: Princeton University Press.
Gray, H. (1858). *Anatomy Descriptive and Surgical*. London: John W. Parker & Sons.

Henrich, J., Heine, S. J., and Norenzayan, A. (2010). 'The Weirdest People in the World?' *Behavioral and Brain Sciences* 33: 61–83.
Hull, D. L. (1986). 'On Human Nature.' *Proceedings of the Biennial Meeting of the Philosophy of Science Association* 2: 3–13.
Hume, D. (1978). *A Treatise of Human Nature*. Oxford: Clarendon Press.
Hume, D. (1995). *David Hume: Essays Moral, Political, and Literary*. Online Library of Liberty: oll.libertyfund.org/titles/704.
Kail, P. (2012). 'The Sceptical Beast in the Beastly Sceptic: Human Nature in Hume.' *Royal Institute of Philosophy Supplement* 70: 219–31.
Lewens, T. (2012a). 'Pheneticism Reconsidered.' *Biology and Philosophy* 27: 159–77.
Lewens, T. (2012b). 'Species, Essence and Explanation.' *Studies in History and Philosophy of Biological and Biomedical Sciences* 43: 751–7.
Lewens, T. (2017). 'Human Nature, Human Culture: The Case of Cultural Evolution.' *Interface Focus* 7. doi: 10.1098/rsfs.2017.0018.
Macauley, W. (2010). *Picturing Knowledge: NASA's Pioneer Plaque, Voyager Record and the History of Interstellar Communication, 1957–1977*. PhD thesis, University of Manchester.
Machery, E. (2008). 'A Plea for Human Nature.' *Philosophical Psychology* 21: 321–9.
Mollon, J. (1989). '"Tho' she kneel'd in that place where they grew…": The Uses and Origins of Primate Colour Vision.' *Journal of Experimental Biology* 146: 21–38.
NASA (2007). 'The Pioneer Missions': www.nasa.gov/centers/ames/missions/archive/pioneer.html. Accessed 7 July 2017.
Pinker, S. (2003). *The Blank Slate: The Modern Denial of Human Nature*. New York: Penguin.
Rivers, W. H. R. (1901). *Reports of the Cambridge Anthropological Expedition to Torres Straits: Physiology and Psychology*, vol. 2. Cambridge: Cambridge University Press.
Sagan, C. (1973). *Cosmic Connection: An Extraterrestrial Perspective*. Cambridge: Cambridge University Press.
Sahlins, M. (2008). *The Western Illusion of Human Nature*. Chicago: Prickly Paradigm Press.
Samuels, R. (2012). 'Science and Human Nature.' *Royal Institute of Philosophy Supplement* 70: 1–28.
Svensson, L., Mullarney, K., Zetterström, D., and Grant, P. (2010). *Collins Bird Guide*. London: Collins.
Thorpe, W. H. (1958). 'The Learning of Song Pattern by Birds with Special Reference to the Chaffinch *Fringilla Coelebs*.' *Ibis* 100: 535–70.
Tooby, J., and Cosmides, L. (1990). 'On the Universality of Human Nature and the Uniqueness of the Individual: The Role of Genetics and Adaptation.' *Journal of Personality* 58: 17–67.
Tooby, J., and Cosmides, L. (1992). 'The Psychological Foundations of Culture.' In J. H. Barkow, L. Cosmides, and J. Tooby (eds), *The Adapted Mind*, 19–36. New York: Oxford University Press.

# 1
# Doubling Down on the Nomological Notion of Human Nature

*Edouard Machery*

## 1.1 Introduction

How to understand the notion of human nature in light of evolutionary biology and genetics?* In recent years I have defended a particular answer to this question, the nomological notion of human nature: 'human nature is the set of properties that humans tend to possess as a result of the evolution of their species' (Machery 2008: 323; see also Machery 2012, 2016a, 2016b, 2016c). In a nutshell, this new account combines two proposals: the 'universality proposal' and the 'evolution proposal'. According to the former, traits that belong to human nature must be typical of human beings; according to the latter, they must have evolved.

This proposal has been extensively criticized (e.g. Lewens 2012, 2015; Powell 2012; Samuels 2012; Ramsey 2013; Downes 2017; Lewis unpublished; Odenbaugh unpublished). The goal of this chapter is to address these criticisms and to improve the nomological notion of human nature.

Here is how I will proceed. In section 1.2, I will review my proposal about the nomological notion of human nature, drawing on Machery (2008, 2016c), and I will lay out the key concerns with this notion. In section 1.3, I will explain what kind of traits belongs to human nature. Section 1.4 defends the universality proposal. Sections 1.5 and 1.6 defend the evolution proposal.

## 1.2 The Nomological Notion of Human Nature

### 1.2.1 A scientifically respectable notion of human nature

Philosophers and scientists agree that the traditional essentialist notion of human nature should be rejected in light of evolutionary biology and genetics (Hull 1986;

---

* I am grateful for Tim Lewens's detailed comments.

Ghiselin 1997; Machery 2008).[1] There are, of course, many properties that all human beings possess: the genetic material of all human beings consists of deoxyribonucleic acid (DNA), blood is made of erythrocytes, leukocytes, and thrombocytes, and so on. However, it is unclear whether there is any property that all and only humans possess— for instance, the genetic material of most organisms is made of DNA (viruses are an exception)—and even if there is, this is a contingent fact. Except perhaps for the relational trait of having a particular genealogy,[2] no trait is essential to human beings *qua* human beings. If it happened that all human beings had brown hair, that would be a contingent fact.

Many (although not everybody—see Downes 2017) also agree that philosophers and scientists should develop an alternative or successor notion of human nature that is (i) useful and (ii) consistent with evolutionary biology and genetics (Machery 2008). The nomological notion of human nature is one of the notions developed in the last ten years in order to meet this challenge (for alternatives, see Griffiths 2009, 2011; Stotz 2010; Samuels 2012; Ramsey 2013; Kronfeldner et al. 2014; Klasios 2016; Odenbaugh unpublished). This notion characterizes human nature as the set of traits that human beings tend to possess as a result of the evolution of their species. The traits that are constitutive of human nature need not be distinctive of humans; they can be shared by humans and other animals. They need not be shared by all human beings; they merely need to be typical. When I refer to the traits acquired by human beings 'as a result of the evolution of their species', I do not merely refer to the traits that have become typical since the emergence of *Homo sapiens*. Any typical trait that evolved in the lineage leading to our species is part of human nature. Human nature is also not meant to distinguish human beings from other animals; this distinction is drawn in genealogical terms. Human nature is not a cause; rather, it is constituted by the outcomes of various causal (evolutionary) processes. Evolutionary processes are understood broadly, and are not limited to selective processes. Finally, this notion is sufficiently bounded: not every trait belongs to human nature. Traits that are not widely shared among human beings are not part of human nature, and traits that are widely shared, but not in virtue of the evolution of our species, are not part of human nature either.

### 1.2.2 *Two noteworthy features of the nomological account*

First, on this view the nature of a species can be thicker or thinner. The more traits conspecifics have in common, the thicker the nature of the relevant species. Some species will have a thicker nature than others. Explaining why the nature of a given species is thick appeals to facts about the evolutionary history of this species, the diversity of its ecological niches, and so on. Human beings' nature seems to be on the thicker side of the continuum of thickness. This fascinating fact about humans is, naturally, contingent, and it is to be explained by looking at the evolutionary history

---

[1] For critical discussion of the exact commitments of pre-Darwinian biology, see e.g. Lennox (2001), Winsor (2003, 2006).

[2] Some even deny that having a genealogy is an essential property of conspecifics (e.g. Slater 2013).

of humans, at the genetic flow between human populations, at the patterns of mate choice, and so on.

Second, as was noted in (Machery 2008: 324), the nomological account of human nature entails that human nature is not fixed: it can change, as human beings evolve. For instance, a dark skin colour may have been part of human nature until recently, since most human beings had a dark skin perhaps as recently as 7,000 years ago (Olalde et al. 2014).

### 1.2.3 What can this notion do for us?

Despite being watered down in some respects, this notion of human nature is far from toothless. First, it provides an explication of the notion of human nature often used in the human behavioural sciences. Consider the following two examples (more in Machery 2016c). Evolutionary biologists and anthropologists Peter Richerson and Robert Boyd describe their overall project in the following terms:

> In the case of ordinary learning, individuals must have some way of weighting the importance of the value of *L* [the trait acquired] that they acquired by imitation against the value that their experience indicates is the best. Do they rely on their experience or on imitation? In the case of biased transmission, individuals must have some criteria of success—do they imitate wealthy individuals? [...] *Ultimately, these are questions about human nature.* The answers must be sought in the long-run processes that govern the interactions of cultural and genetic evolution in our species.   (2005: 392–3; emphasis added)

In an article published in *Science*, economist Herb Gintis describes some recent modelling of cooperation in human beings (see Hermann et al. 2008):

> The standard view holds that *human nature* has a private side in which we interact morally with a small circle of intimates and a public side in which we behave as selfish maximizers. Herrmann *et al.* suggest that most individuals have a deep reservoir of behaviors and mores that can be exhibited in the most impersonal interactions with unrelated others. This reservoir of moral predispositions is based on an innate prosociality that is a product of our evolution as a species, as well as the uniquely human capacity to internalize norms of social behavior. Both forces predispose individuals to behave morally even when this conflicts with their material interests.
> (Gintis 2008: 1345; emphasis added)

What do Richerson, Boyd, and Gintis mean by 'human nature'? How are we to understand this notion in light of evolutionary biology and genetics? The nomological notion of human nature provides an answer.

Second, the nomological notion of human nature fulfils many of the functions the traditional notion of human nature was meant to fulfil (as evidenced by the role of appeals to human nature in philosophical, political, or moral texts; see e.g. Antony 1998), while leaving unfulfilled those that ought not to be fulfilled (Machery 2016c). The nomological notion of human nature fulfils a descriptive, explanatory, and limitative function. Identifying the components of human nature amounts to describing what human beings are like.[3] One can also explain people's behaviour by saying that it is part of human nature: while human nature is not a cause according to the nomological notion, the notion of human nature is still a causal-explanatory notion. Classifying a trait as part of human nature ('It's human nature to lie') is to endorse a particular causal-explanatory schema for this trait—one that asserts that this trait has an evolutionary aetiology—and to insist on the inappropriateness of attempts to provide a causal explanation of this trait by non-evolutionary schemas alone. It can also fulfil some version of the limitative function: human nature puts, in some sense, limits on what human beings can do. Because many traits that constitute human nature have a long phylogeny, they are likely to be insensitive to education and cultural factors, the kind of factors we know how to manipulate. Other traits are likely to be canalized against variation in the kind of factors we can influence. And modifying traits that belong to human nature may have widespread unexpected consequences for other aspects of a human life. That the nomological notion of human nature can fulfil these functions has been defended at length elsewhere (Machery 2016c), and will be taken for granted in what follows.

### 1.2.4 Objections

The nomological notion of human nature has been submitted to severe criticisms, some of which have been addressed in more or less detail elsewhere (Machery 2012, 2016c). The remainder of this chapter will address the criticisms that have not yet been addressed, or those that have not been addressed with the required detail. To anticipate, here are the central objections:

- The notion of trait is left unspecified.
- The notion of typicality is too vague.
- Only distinctive traits should belong to human nature.
- There is no reason to limit human nature to typical traits.
- Every trait is in some sense the result of evolution.
- One cannot distinguish the traits that people tend to possess as a result of evolution from those that people tend to possess because of culture.

## 1.3 What Kind of Trait Belongs to Human Nature?

My original formulation of the nomological account (Machery 2008) says little about what kind of trait is constitutive of human nature, and this lack of precision obscures the nomological account. This section clarifies the issue.

---

[3] The need to fulfil the descriptive function explains why the nomological notion of human nature does not include only adaptations (Klasios 2016).

## 1.3.1 Dispositions

One may first ask whether, according to the nomological notion of human nature, the traits that are part of human nature are meant to be exclusively categorical or include dispositions too. How to draw the contrast between categorical properties and disposition remains controversial in philosophy, but for our purposes an intuitive understanding suffices. 'Flat' and 'square' are categorical properties. 'Being flammable' and 'being soluble' are dispositions. An object (e.g. a piece of sugar) has a disposition (soluble) when it would possess a categorical property (e.g. dissolved) were it subjected to some kind of treatment (e.g. were it put in a glass of water).

Human nature cannot include only categorical properties. Human beings are disgusted, angry, happy, jealous, and ashamed by nature, but they aren't continuously disgusted, angry, and so on. At any given time, a small proportion of human beings may feel any of these emotions. This suggests that when we say that disgust is part of human nature, we really mean that the disposition to feel disgust under some conditions (to be discovered by empirical research) is part of human nature. While it's not typical of human beings to be angry, it is typical of them to be disposed to be angry.

A problem is lurking here. Including dispositions within human nature threatens to undermine the role of typicality in distinguishing the traits that are part of human nature from the traits that aren't. Consider a trait that is not typical, such as speaking French. Most human beings have the disposition to speak French: they would speak French if they were taught French. Since the disposition to speak French is not a product of evolution (for discussion, see sections 1.5 and 1.6), the disposition to speak French still does not belong to human nature according to the nomological account of human nature; but now only the evolution proposal seems to play a role in sorting out the traits that are part of human nature from the traits that aren't.

This difficulty can be alleviated by introducing a distinction between capacities and dispositions. Most people have a capacity to speak French, but they are not disposed to speak it. The intuitive difference between the former and the latter is that to speak French, people would have to change: they would need to learn French. By contrast, people are disposed to feel ashamed: without changing, they feel shame when they are dishonest.[4]

## 1.3.2 Broadening the nomological account

Some traits are not possessed by human beings, but seem nonetheless to be good candidates for being part of human nature. First, there are those traits that human beings don't possess *by themselves*. I gave the example of biparentality in Machery (2008), but no individual human being can possess this trait; rather, it is a trait of pairs of human beings.[5] More generally, we may want to ensure that the explicated notion of human nature leaves room for traits possessed by typical groups of human beings,

---

[4] Spelling out this distinction in more detail is quite difficult (the metaphysics of dispositions is itself unclear), but our intuitive grasp is sufficient for present purposes.

[5] Thanks to Grant Ramsey for noting this point.

including reproductive pairs, families, friends, and so forth. Such traits may include dominance and prestige hierarchy (e.g. Henrich and Gil-White 2001), ethnic markers (McElreath et al. 2003), and the size of typical human groups (e.g. Hill and Dunbar 2003). Fortunately, a minimal modification of the nomological account is sufficient for leaving room for such traits: human nature includes the traits that human beings or typical human groups tend to possess as a result of evolution. The qualification 'typical' in 'typical human groups' is meant to exclude groups of human beings that are found in only some cultures (e.g. baseball teams). That is, we identify those groups that are found in most cultures (e.g. familial units and groups of friends, but not baseball teams), and we then identify the properties most of these groups share. Biparentality and dominance hierarchy are now part of human nature according to this minimally revised notion.

Other traits are possessed by human beings, but not throughout their life. Consider, for instance, the traits that are characteristic of children—say, being unable to ascribe false beliefs explicitly before the age of 4 (Liu et al. 2008)—or of older people—say, an increased risk aversion. Because most human beings are neither children nor old people, it is not true that being unable to ascribe beliefs or increased risk aversion are traits that human beings tend to possess. On the other hand, it is natural to think that describing what human beings are like involves describing what they are like at the various life stages that constitute the life history of human beings. Again, a minimal modification of the nomological account is sufficient for leaving room for such traits: human nature includes the traits that human beings tend to possess either throughout their life or during a stage of a typical human life history because of the evolution of their species. So, the species-typical traits of children younger than 4 are part of human nature, as is the species-typical increased risk aversion of older human beings (if it is really species-typical).

To summarize then: human nature is the set of traits—including dispositions—that human beings, either throughout their life or during a stage of a typical human life history, or typical human groups tend to possess because of the evolution of their species.

### 1.3.3 From physiology to behaviour

The notion of human nature is often associated with psychological traits, in part because of its association with psychological research; but its scope is broader according to the nomological notion. Human nature includes physiological and anatomical traits, as well as psychological and behavioural traits, as suggested by the examples given in Machery (2008). It's part of human nature that the cortex possesses various lobes, that the liver is made of hepatocytes, and so on.

A behavioural trait like cooking is also part of human nature according to the nomological account. Cooking, or at least the disposition to cook, is typical of human beings, and its distribution has an evolutionary explanation (more about the role of evolutionary explanations in section 1.5). People cook because, due to the importance of fire during human evolution, the gut shrank and was less efficient at extracting calories from ingested food (Wrangham 2009).

## 1.4 Why are Human-Nature Traits Typical?

The nomological account of human nature singles out traits that are *typical* of human beings. This idea has been criticized on various grounds, and it calls for a number of clarifications.

### 1.4.1 The vagueness of typicality

One could be concerned that the notion of typicality at play in the nomological notion of human nature is too vague for it to do any theoretical work. If we can't specify more precisely what proportion of human beings must possess a given trait for it to be included in human nature, then it will often be unclear whether or not a trait belongs to human nature.

This objection is not threatening, and I propose to bite the bullet. The nomological account does leave the notion of human nature vague—on this view, it is indeterminate whether *some* traits belong to human nature—but this is exactly as it should be. The notion of human nature that is explicated by the nomological account of human nature is itself vague, and the nomological account does not attempt to mitigate this vagueness because it does not prevent the notion of human nature from fulfilling the roles it should fulfil (on these roles, see Machery 2016c). Sharpening it further would then be artificial.

### 1.4.2 The relativity of human nature

The traits that are part of human nature change over time (as was noticed above), since human beings evolve. As a result, the notion of human nature is *relative* to a particular specification of the class of human beings. Human nature will not be the same when it is relative to our contemporaries, to human beings in the last few millennia, or to all members of the species *Homo sapiens*. One may think that this relativity of human nature is a problem, but it is not.

According to the nomological account, characterizing human nature amounts to describing what humans are like, thereby fulfilling the descriptive function of the notion of human nature. Since what human beings are like changes (as human beings evolve), the notion of human nature can only fulfil the descriptive function if it is relativized to a particular class of human beings. Far from being a bug, the (explicit or implicit) relativity of the notion of human nature to a particular specification of the class of human beings is a feature of the nomological account.[6]

### 1.4.3 Typicality versus distinctiveness

The traits that constitute human nature according to the nomological account are not necessarily distinctive of human beings; they may be shared with other animals. Examples include bipedalism, incest avoidance, fear, and shame—traits that are not

---

[6] I am grateful to Liam Kofi Bright for highlighting this consequence of the nomological account.

distinctive of human beings. One may object that human nature should, on the contrary, include only distinctive traits of human beings. Thus, Odenbaugh writes the following against both Ramsey and myself:

Theories of human nature can give up essential properties but they cannot give up distinctive properties. After all, we want to explain what makes us distinctively human. Machery makes the same mistake since he includes traits like bipedalism as part of our nature. But other species are bipedal; *Struthio camelus* otherwise known as the ostrich is. Bipedalism is not distinctive of our species. (Odenbaugh unpublished)

Similarly, Buller (2005: 420) insists that the concept of 'human nature' refers only to 'what distinguishes humans from the other animals on the planet' (for discussion, see also Machery and Barrett 2006).

There is no doubt that the notion of human nature that contemporary accounts of human nature attempt to replace was meant to distinguish human beings from other animals. The Boolean function of traits thought to constitute human nature, if not each of these traits, was meant to be possessed by human beings and only by human beings. Even if one relaxes the requirement that the traits constitutive of human nature be possessed by all human beings, as Odenbaugh seems inclined to do, one may think that only traits that are distinctive of human beings should be part of human nature.

There are two lines of response to this point. First, the nomological account of human nature is in part meant to explicate the notion of human nature used in the evolutionary behavioural sciences, including evolutionary psychology and human behavioural ecology, when they describe their research as an investigation of human nature. And evolutionary behavioural scientists do not focus exclusively on traits that are distinctive of human beings. Evolutionary behavioural scientists investigate, among other things, species-typical traits, whether or not they are shared by other species. Building distinctiveness into the explication of the notion of human nature would rule out staples of evolutionary research, such as those concerning face recognition (Duchaine et al. 2001) and kin-directed altruism (Kurland and Gaulin 2005), that are not unique to humans. Admittedly, some evolutionary behavioural scientists may highlight traits that are distinctive of human beings when they refer to human nature, but their investigative practices are not limited to these traits.

The second line of response to Odenbaugh's criticism focuses not on the task of explicating the notion of human nature used by evolutionary behavioural scientists, but on the descriptive function of the notion of human nature. A satisfying notion of human nature must be such that describing human nature tells us what human beings are like. This function can be understood in two different ways: (1) what human beings are distinctively like (in contrast to all other animals) or (2) what human beings are in general like. Only the first reading builds distinctiveness into the notion of human nature. One can't choose between readings (1) and (2) without a whiff of stipulation: it is unclear what convincing arguments could favour one of them. If this is correct, then embracing the second reading, as the nomological account does, can't be held against it.

### 1.4.4 Typicality versus demographics-specific traits

Building the notion of typicality into the notion of human nature, as the nomological account does (see also Machery and Barrett 2006), results in excluding from human nature the traits that only some groups of human beings possess. (I will call these traits 'group-specific traits' to contrast them to species-typical traits.) If a trait is typical of women only (for example, lactation and menopause) or of men only (for example, facial hair and Adam's apple), then it is not part of human nature on this view. If a trait is typical of French people only (wearing a beret?), then it is not part of human nature. One may object to this consequence of building typicality into the notion of human nature on at least four different grounds. First, for what it's worth, it may be counterintuitive to exclude from human nature the typical traits of women or of men, and more generally of particular demographics. Second, excluding traits that are typical of, say, men or women would undermine the nomological account's ability to successfully explicate the notion of human nature as used by evolutionary behavioural scientists. As Lewens puts it: ' "male philandering" is just the sort of trait that evolutionary psychologists are likely to regard as part of human nature, and yet only half of humans are male' (2012: 463–4). Third, this consequence of typicality prevents the nomological account of human nature from satisfactorily fulfilling the descriptive function. Ramsey writes:

> by requiring possession by the majority of humans, one loses many traits characteristic of humans. Any traits (psychological, behavioral, morphological) that are sexually dimorphic or, say, exhibited only by a particular ethnic group, will be excluded. Viviparity, lactation, and menopause, for example, are no part of human nature. (2013: 985)

Finally, one could object that there is no justification in the biological sciences for singling out typical traits over non-typical traits of human beings. Ramsey continues his criticism of the nomological account along these lines:

> But why does belonging to almost all humans make it an important class for biology? Why is this similarity more important than diversity? Is it not a biologically interesting feature of human beings that the females undergo menopause? Furthermore, why should we presume that it is the sameness across individuals that is of interest to scientists, and not variation? (2013: 986)

I address these four objections in turn. Whether or not it is intuitive to exclude from human nature traits that are not species-typical should ultimately carry little weight in the present debate. Furthermore, it strikes me as perfectly intuitive to distinguish human nature from the nature of, say, men and women. Evolved traits that are only typical of women—lactation and menopause, for instance—belong to the nature of women, not to human nature. The second objection is more interesting, but I propose to bite the bullet. The nomological account will result in some minor revisions of what counts as an inquiry into human nature.

The third objection is important, but it too fails. It is not the case that a trait is characteristic of human beings—and thus belongs to human nature (assuming a proper notion of human nature must fulfil the descriptive function)—just because some human being happens to possess it. For instance, the scar above my left knee is not characteristic of human beings; neither are the distinctive facial features that run in the Machery family. So, someone like Ramsey who wants to uphold the notion of human nature while rejecting the universality proposal must single out some demographics in a non-arbitrary way in order to identify those group-specific traits that must be included in human nature. If it turns out to be impossible to single out some demographics, then anybody who wants to uphold the notion of human nature should accept the universality proposal.

Ramsey does not explain clearly how to distinguish the group-specific traits that should be included in human nature from those that should not. He gives two examples—traits of men and women, and ethnic traits—but these examples are unhelpful. It is not clear what he has in mind by traits 'exhibited only by a particular ethnic group': racial traits or something else? In any case, human beings do not form ethnic groups at a single level; rather, groups identified at one level partition into smaller ethnic groups, which themselves partition into further ethnic groups. (And divisions of human beings into ethnic groups often cross-cut each other.) So, if ethnic identity is an appropriate demographic for identifying the group-specific traits that should belong to human nature, then traits specific of Sicilians (or Sardinians or Corsicans) should be included into human nature. But if those are, why not the facial features that run in the Machery family? The distinction between the former and the latter would seem to be artificial.

It may be tempting, although misguided, to respond that only biological demographics should be used to identify the group-specific traits that belong to human nature. I will mention only two problems with this response. First, the notion of a biological demographic is not clear. Second, supposing this notion can be clarified, families are probably biological groups, and this proposal would not exclude the distinctive facial features that run in the Machery family.

Perhaps Ramsey could claim that only demographics that partition all the other demographic partitions should be used to identify the group-specific traits that belong to human nature. In all cultures, all ethnic groups, all families, all languages groups, and so on there are men and women, so gender or sex—I'll bracket the difficult question of determining which of these two demographics would be appropriate—could be used to identify group-specific traits in a non-arbitrary manner.[7] There are other, less obvious demographics of this ilk: psychopaths are found in all cultures, all ethnic groups, all families (provided the latter are large enough), and so on. Belonging to the Machery family is excluded by this criterion. The typical properties of ethnic groups (whatever those are) would also not be included in human nature (in contrast to Ramsey's proposal quoted above), since having a particular ethnic identity does not partition all the other demographic partitions. Ramsey could probably accept this consequence if the claim under consideration allowed him to reject the

---

[7] Groups whose definition appeals to gender or sex (e.g. male-only golf clubs) won't be partitioned by means of gender or sex, but this is easily taken care of by restricting the proposed principle to groups not defined by appealing to gender or sex.

universality proposal. The trouble with this claim, however, is that the traits typical of extremely rare developmental variants (e.g. Williams–Beuren syndrome) end up being part of human nature.

I now turn to the final and most important argument, expressed by Ramsey's second quotation (see also Downes 2017). As Ramsey rightly notes, evolutionary thinking does not single out typicality as being particularly important for characterizing a species biologically. Sober (1980) helpfully contrasts natural-state models for understanding variation (illustrated on his view by Aristotle's biology) and population-thinking models. The former understand variation as deviation from the natural state of a species. Typicality is obviously a central concept in these models, since the typicality of a trait is at least evidence for it being the natural state of the species. Population thinking explains variation differently: variation within a population is explained by further variation subjected to various forces. As Sober puts it, focusing on the contrast between Quetelet and Galton:

> For Quetelet, and for typologists generally, variability does not explain anything. Rather it is something to be explained or explained away. Quetelet posited a process in which uniformity gives rise to diversity; a single prototype—the average man—is mapped onto a variable resulting population. Galton, on the other hand, explained diversity in terms of an earlier diversity and constructed the mathematical tools to make this kind of analysis possible. (Sober 1980: 370)

In population-thinking accounts of variation, typicality, which is acknowledged to exist, is not of greater significance than variation.

It is important to see that the nomological account is not committed to the view that typicality is evidence for the natural dispositions of a species, and thus does not endorse the natural-state model. I made this point in Machery (2008) by contrasting my approach to Aristotle's. And the nomological account is consistent with population thinking: similarities and differences coexist within populations, and both need to be explained by appealing to the changes in the distribution of traits across generations.[8]

Be that as it may, I still have not answered Ramsey's challenge: why focus on similarity given that it has no particular explanatory and evidential significance in evolutionary biology? While typicality has lost its explanatory and evidential significance in a population-thinking framework, it remains an explanandum: one still needs to explain why similarities are found across environments, cultures, and so forth, particularly when conspecifics, such as human beings, happen to have many traits in common (when their nature is thick, to use the terminology introduced earlier). Of course, typicality is not the only explanandum for evolutionary biologists—variation also matters—but that there are other explananda does not mean that typicality is not to be explained. So, *pace* Ramsey, embracing the nomological notion of human nature

---

[8] To be crystal clear, the nomological account does not assume that only similarities are the products of evolution. Evolution results in similarities and differences within populations.

does not amount to presuming 'that it is the sameness across individuals that is of interest to scientists, *and not variation*' (my emphasis). And to say that it is an explanandum is not to say that it is a 'more important' explanandum than diversity, contrary to what Ramsey suggests. So, the reason why typicality is built into the notion of human nature is not that typicality has a particular explanatory or evidential place in evolutionary biology, but that similarities are one of the explananda in biology, and that we need a notion to single out this explanandum: the notion of human nature is that notion in the case of humans.

A critic may concede that the nomological notion of human nature is not committed to the natural-state model, to the view that only similarities across human beings are to be explained, or to the view that explaining similarity is more important than explaining diversity. Rather, the problem with the nomological notion is simply that typicality is 'singled out'. From an evolutionary point of view, the critic goes on, similarities and differences between conspecifics are on a par; they are particular cases of the distribution of traits within populations.[9] Some traits are widely distributed, others not. What needs to be explained, then, is the distribution of traits within populations; and from this perspective, there is no justification for distinguishing particular types of distribution, contrary to what the nomological notion of human nature proposes to do.

This criticism has a point: the phenomenon of interest for population thinking, as embodied for instance in population genetics, is the distribution of traits in populations, and there is no justification from this point of view for distinguishing typical traits from traits that are not typical. Hence, the nomological notion of human nature is not justified by considerations drawn from population thinking. But this does not mean that this notion is inconsistent with population thinking: there are typical traits; their typicality must be explained; and, from perspectives other than population thinking, they can justifiably be singled out. Justification can be found in disciplines outside biology that single out typical traits, such as cognitive science, linguistics, and some research traditions in anthropology. It can also be found in biology: population thinking is only one of the explanatory styles in biology; to give only two examples relevant to the study of human nature, explanatory styles in physiology and medicine are distinct from population thinking (e.g. Wachbroit 1994). It can even be found in evolutionary biology: as has been often discussed by philosophers of biology and evolutionary developmental biologists themselves, evo-devo is typological (e.g. Amundson 2005). And it is not surprising, given that they developed at the intersection of various other disciplines, that at least some of the concepts at work in the evolutionary behavioural sciences—the home of the notion of human nature explicated by the nomological notion—are not grounded in population thinking (although, again, they need not be inconsistent with it).

---

[9] Tim Lewens suggested this to me in conversation.

## 1.5 Why are Human-Nature Traits Evolved? Part I

Many philosophers have objected that it is not possible to distinguish among the typical traits of human beings those that they tend to possess because of the evolution of their species from those they possess for other reasons. As noted above, I call this proposed distinction the 'evolution proposal'. There are several versions of this general objection. In this section, I focus on Ramsey's version:

> Any organismic trait is going to be due to both heritable features of the organism as well as the particular environmental features the organism happens to encounter during its life. Some of these environmental features could be counted as instances of 'enculturation' or 'social learning', but the fact that such environmental features are present in the organism's life history does not mean that we can point to properties as being 'exclusively due' to these environmental inputs. The innate–acquired dichotomy has been long challenged [...] and I see no way to make Machery's distinction without a futile attempt at reifying this problematic dichotomy. (2013: 986)

Ramsey interprets the evolution proposal in developmental terms (see also Powell 2012: 488–9). On his view, one could draw this distinction if one could identify traits whose developmental aetiology is causally influenced only by processes such as enculturation and social learning. Since he doubts that it makes sense to search for those traits, he concludes that the evolution proposal cannot be drawn. Further, he suggests that attempting to draw this distinction amounts to embracing the distinction between innate and acquired traits.

Before responding to this objection, a few clarifications are in order. I will argue that Ramsey misconstrues the evolution proposal, but, in all fairness, his interpretation may have been prompted by the claim that 'saying that a given trait, say [...] outgroup bias, belongs to human nature is to say that some kinds of explanation for the occurrence of this trait among humans are inappropriate. Particularly, this is to reject any explanation to the effect that its occurrence is exclusively due to enculturation or to social learning' (Machery 2008: 326). While this claim does not say that the development of any trait is due only to enculturation, it could perhaps be read as claiming this (more on how to understand it properly below). Second, for the record, I agree with Ramsey that the distinction between innate and acquired traits is bankrupt, and cannot be reconstructed, and I have indeed made this claim in print in a few places (Griffiths et al. 2009; Griffiths and Machery 2008; Linquist et al. 2011; Machery 2016b). More important, the evolution proposal does not require that one be able to distinguish innate from acquired traits. Let's now see why.

Ramsey is right that no trait develops exclusively under the influence of developmental processes such as individual learning or social learning, and no account of human nature should entail that some traits do. Fortunately, the nomological account does not. The nomological account is not about development, but about the distribution of traits within a population; and it is not about the factors that influence the distribution of a trait, but about the factors that figure in the explanation of its distribution. Thus, the nomological account requires that one be able to distinguish those traits whose distribution is explained by appealing only to individual or social learning from other traits, not traits whose development is only influenced by individual or social learning from other traits. The difference may seem subtle, but it is important. Even if the development of all traits is influenced by a mixture of causes, and never only by individual or social learning, the explanation of the distribution of some traits among human beings—that is, why some traits are typical of human beings—but not the explanation of the distribution of other traits, appeals only to environmental factors such as cultural and social learning.

First of all, that the explanation of the distribution of some traits among human beings appeals only to environmental factors such as cultural and social learning is intuitive. Consider, for instance, the explanation of why a very large majority of Americans know the lyrics of the 'Star-Spangled Banner'. A satisfying explanation of the distribution of this trait would appeal to the fact that it is the American national anthem, that it is taught at school, and that most American children go to school. So, while the process by which an American comes to learn the 'Star-Spangled Banner' is clearly very complex and involves various kinds of causes, whose description falls in the purview of different sciences (e.g. neuroscience, psychology, sociology, and history), it is still the case that, intuitively, we satisfyingly explain the distribution of this trait by appealing to cultural and social learning, together with facts about Americans' social environment; facts about the evolution of hearing and so on are not explanatory.

Of course, we should do better and provide an argument for this intuitive claim. Two elements are needed: a claim about explanation and a claim about what makes a candidate explanans a genuine explanans. First, then, explanation. I focus only on why-explanation, the relevant type of explanation for spelling out the evolution proposal. On my view, explanation is always contrastive (van Fraassen 1980; Lipton 1990). One explains why $e$ (an event, a generalization, a law of nature, and so on) rather than $e^*$ ($e$'s foil) holds. Depending on which contrast is drawn, different explanantia may be called for. One would not explain in the same way why the walls of an apartment have been painted blue rather than red, and why they have been painted rather than covered with wallpaper. The contrastive nature of explanation is often left implicit, but the relevant contrast is part of a common background of presuppositions. Now, on to the second claim: assessing candidate explanantia. On my view, the explanatory quality of candidate explanantia bearing on an explanatory contrast depends in part on their specificity. A candidate explanans is specific with respect to a contrast between $e$ and its foil $e^*$ if and only if it explains why $e$ *rather than* $e^*$ holds. Candidate explanantia that are not specific—they in some sense explain why $e$ holds, but not why $e$ rather than $e^*$ holds—fail to provide a genuine explanation. For instance, the fact that Americans can hear and enjoy music fails to explain why they know the 'Star-Spangled Banner', because it fails to explain why they know the 'Star-Spangled Banner' rather than some other song. For the same reason, adding the fact that Americans can hear and enjoy music to the fact that the 'Star-Spangled Banner' is the American national anthem does not improve the quality of the explanation of the fact that Americans know the 'Star-Spangled Banner'. In fact, it makes it worse, since part of the explanans is now not explanatory (since it is not specific enough).

Applying the contrastive account of explanation to the issue at hand, the nomological account of human nature is committed to the legitimacy of the following distinction. The typicality of some traits is satisfyingly explained by appealing to individual or social learning, and evolutionary information is not explanatory. By contrast, the typicality of other traits can be explained by appealing to evolutionary information, although they can also be explained by ontogenetic information (including individual or social learning). I use 'evolutionary information' as a cover term for the class of factors that explain the evolution of traits. This includes phylogenetic facts, natural selection, drift, niche construction, and developmental constraints. As the nature of these factors is controversial among evolutionary biologists and philosophers of biology, I'll remain neutral about what constitutes evolutionary information.

Now, to see why the distinction just proposed is legitimate, remember that explanation is contrastive. If most human beings have learned the rules of football (the real one, the beautiful game), then we need to explain why the rules of football rather than, say, the rules of rugby or American football are widely known. If most human beings have learned by individual learning that the sky is blue, then we need to explain why they have formed this perceptual belief rather than, say, the belief that it is green. Remember now that genuine explanantia are specific: information that fails to explain why $e$ rather than its foil $e^*$ holds is not explanatory. Some contrasts are such that only information about individual or social learning selectively explains why some traits rather than their foils are typical. Consider this (hypothesized) explanandum: the rules of football rather than, say, the rules of rugby are widely known. Information about the evolution of the human grasp of norms, the human capacity to play games, and so forth is not specific to the contrast between the rules of football and those of rugby, in contrast to information relevant to social learning. A proper explanation of this contrast is historical or sociological. Consider now the following explanandum: human beings have trichromatic colour vision rather than monochromatic, black-and-white vision. While this explanandum has an ontogenetic explanation (involving human beings' genes and a complex set of developmental facts), it also has an evolutionary explanation: information about the phylogeny of human beings and the selective environment of some of our primate ancestors explains why human beings have trichromatic colour vision rather than monochromatic, black-and-white vision (e.g. Shyue et al. 1995).

The appeal to contrastive explanation to flesh out the nomological account of human nature casts some light on the following passage: 'evolutionary processes causally contribute to the existence of any trait that is common among humans. But only some of these traits can be explained by reference to evolutionary processes. That is, only some of them are the object of ultimate explanations' (Machery 2008: 327). The key point here is that only some traits are such that their distribution is satisfyingly explained by appealing to evolutionary information (although their development can also be explained by ontogenetic processes); other traits cannot be so explained. Why? To repeat, because evolutionary information would not be specific enough, and thus would not provide a genuine explanation. For instance, if a few years ago most human beings had learned Psy's song 'Gangnam Style', one would not provide a proper explanation of this explanandum by appealing to the evolution of the larynx and the vocal chords, or the evolution of the human musical sensitivity (e.g. Mithen 2005) since evolutionary information would not be specific to the contrast between Psy's song 'Gangnam Style' and Carly Rae Jepsen's 'Call Me Maybe'.

It must be noted that in the passage just quoted, I appealed to Mayr's (1961) distinction between ultimate and proximate causes; but I am now inclined to make the point without appealing to this distinction. It is notoriously unclear, and it has been interpreted in several ways over the years. Furthermore, at least some of these interpretations (particularly those that exclude from evolutionary explanations facts about development) are controversial (e.g. Laland et al. 2011). The distinction between evolutionary and ontogenetic information should by contrast be uncontroversial, provided one makes it clear that the former can involve developmental facts. Evolutionary information explains why a trait evolved the way it did; developmental information explains why it developed the way it did.

One could perhaps object that information about the evolution of human musical sensitivity might well explain why in the fictional scenario presented above 'Gangnam Style' rather than 'Call Me Maybe' would have become known all over the world. Indeed, an important tradition in the evolutionary behavioural sciences—the epidemiological model of cultural evolution (Sperber 1996; Sperber and Hirschfeld 2004)—explains the typicality of cultural variants in precisely this way. Some cultural variants happen to be widespread because they 'fit' the evolved structure of the human mind. Embracing this type of explanation, Boyer (2001) has argued that representations of gods are typically 'minimally surprising': they violate a few commonsense expectations, while being otherwise consistent with folk physics, folk psychology, and so on. More recently, Wengrow (2013) has argued that representations of mythical monsters are found all over the world because they too are minimally surprising.

In response, it must be conceded that information about the evolution of the human musical sensitivity might well explain the distribution of 'Gangnam Style'. The crucial point, however, is that it might not, which shows the legitimacy of the distinction between different kinds of typical traits upon which the nomological notion of human nature is built. Furthermore, if the typicality of 'Gangnam Style' were explained by appealing to information about the evolution of human musical sensitivity, then it would be part of human nature, exactly as it is part of human nature to believe in a certain type of god or to have a certain type of representation of monsters. According to the nomological account of human nature, the fact that representations of gods or monsters are culturally transmitted is compatible with them (or, more precisely, compatible with a particular type of god or monster representation) being part of human nature; the only thing that is required for this inclusion are typicality and the explanatory relevance of evolutionary information.

A second objection focuses on the contrastive nature of explanation. It seems to entail that whether a trait is part of human nature depends on the contrast drawn, and the same trait may be part of human nature for some contrasts, but not for others. For instance, suppose that human beings typically represent out-groups as disgusting. Suppose also that, probably contrary to fact, people typically represent out-group members as being stinky (one of the cues of disgust). The stereotype of out-group members as stinky could fail to be part of human nature when it is contrasted with the stereotype of carrying germs (another cue of disgust); it could be part of human nature when it is contrasted with the stereotype of out-group members as fierce. This would be the case if information about cultural transmission but not evolutionary information explained the first contrast, while information about the evolved nature of the mind explained the second contrast (as proposed by Kelly 2011).

This consequence of appealing to a contrastive theory of explanation may seem strange, but it is exactly as it should be. Different contrasts correspond to different ways of individuating a trait. Evolutionary information may be relevant when a trait is individuated one way, and irrelevant when it is individuated another way. And there is nothing puzzling if a trait is part of human nature when it is individuated one way, while it is not part of human nature when it is individuated differently.

A final objection notes that pragmatic, contextual factors determine what counts as a successful explanation, and then remarks that it would be strange if what counts as part of human nature varied across pragmatic contexts. In response, I grant that explanation is at least partly pragmatic, but insist that this is not an issue for the account proposed here. The pragmatic context determines which contrast must be explained, which type of explanation is expected (whether it is a response to a why- or how-question), and how much detail is required to produce a successful explanation. As we have seen, the contrastive nature of explanation does not raise any serious challenge: in different contexts, different contrasts, and thus different ways of individuating the relevant trait, are considered; only why-questions are relevant here; and to assess whether a trait is part of the human nature, we need to consider the explanatory contexts where more information is expected to be provided.

Wrapping up, the nomological notion of human nature is not committed to unacceptable claims about the development of human traits. It does not claim that the development of some traits is only influenced by individual or social learning; it acknowledges the complexity of the ontogenetic process; and it is not committed to the untenable distinction between innate and acquired traits. But it relies on the following distinction: to explain why some traits are typical, one can appeal to evolutionary information; to explain why other traits are typical, one can only appeal to individual or social learning. The former, but not the latter, are part of human nature. The contrastive nature of explanation and the selectivity of good explanantia explain why this distinction is defensible.

## 1.6 Why are Human-Nature Traits Evolved? Part II

I now turn to the second criticism of the evolution proposal, which is due to Lewens (2012, 2015). His criticism is multi-pronged. First, he notes that gene–culture coevolution theorists treat learning and cultural processes as evolutionary, which undermines the contrast between being explained by evolutionary information and being explained by social learning—the contrast that Machery (2008) and the previous section rely on:

> Machery's thought appears to be that social learning is not itself an evolutionary process, hence if social learning is the only process responsible for the occurrence of some trait, the trait does not count as a part of human nature. To make this argument good one needs some reason for denying that evolutionary processes in general include cultural evolutionary processes. (Lewens 2012: 464)

He then notes that even if the evolution proposal could be drawn, including in human nature only those traits that can be explained by means of evolutionary information seems unprincipled:

> to the extent that 'human nature' is supposed to name those traits that are widely distributed among our species, and perhaps even those traits whose development is very robust, or hard to evade, it's not clear why we should rule out an explanation for their persistence that looks solely to enculturation or social learning. (Lewens 2012: 464–5)

Finally, Lewens holds that the evolutionary proposal is not only unprincipled, it is also theoretically dubious since it distinguishes traits that should not be distinguished:

> Machery's proposal to equate human nature only with those elements of our common makeup that can be understood as modifications of earlier more ancient traits threatens to draw a theoretically dubious distinction between different cognitive traits, all of which are produced via social learning. (2012: 465)

In effect, then, Lewens proposes to reject the evolution proposal: on his view, the process by which a trait has become typical is irrelevant for identifying the components of human nature. (We've already seen that he is also sceptical of the typicality proposal.)

I will address these three criticisms in turn. First, then, the evolutionary nature of social learning. The criticism here is that traits that are acquired by social learning (e.g. various skills or techniques) have an evolutionary history too; they are the products of cultural evolution. Folk tales are acquired by social learning, but evolve from generation to generation. Thus, when the distribution of typical traits can be explained by social learning, it can also be explained by means of evolutionary information, and there is no trait that is explained by social learning and is not explained by evolutionary information, contrary to what was assumed in the previous section. For the sake of the argument, I will concede that cultural change is at least often properly thought of as an evolutionary process, although I should note that this is controversial (for discussion, see e.g. Lewens 2015; Ramsey and De Block 2017).

The first thing to say in response is that if this criticism really showed that socially learned traits that are typical belonged to human nature, many typical traits would still not be part of human nature, namely, the traits that are acquired by individual learning. 'Gangnam Style' would be in, the belief that sky is blue out. The notion of human nature would not be trivialized. Furthermore, one of the goals of the nomological account is to explicate the notion of human nature used by evolutionary behavioural scientists. These would be disinclined to include fads in human nature just because they happened to have become widespread. For our explication to be sufficiently similar to the explicatum (one of the criteria to assess explication in Carnap 1950; Machery 2017), we must exclude traits that have become typical due to purely cultural evolutionary processes.

Lewens would perhaps respond that in this case the explication of the notion of human nature must correct scientists' use of this notion. He could appeal to his second criticism to motivate this correction: it would be unprincipled to exclude purely cultural evolutionary explanations from the type of explanations used to delineate what belongs to human nature. There is without question a whiff of stipulation in this exclusion, but it is not arbitrary. It can be justified by appealing to some of the functions that the notion of human nature should fulfil (on these functions, see Machery 2016c). As we have seen in section 1.2, the notion of human nature has an explanatory function. According to the nomological account, the notion of human nature fulfils this explanatory function because to say that people do $x$ or are $y$ by nature is to say that people doing $x$ or being $y$ is correctly explained by means of a particular explanatory sketch (Machery 2016c). Including different types of evolutionary explanation—cultural evolutionary explanations, in addition to explanations that involve genetic changes or, more broadly, changes in the developmental programmes of conspecifics—would make the explanatory sketch even more abstract and general, and would weaken the explanatory significance of appeals to human nature.

What I have said so far is sufficient to address Lewens's third criticism: it is unclear why distinguishing traits whose distribution can be explained by means of evolutionary information from those that can't would be 'theoretically dubious'; they are distinguished because they call for different explanations. The case for excluding purely cultural explanations is the same. Distinguishing evolved traits whose distribution is explained solely in cultural-evolutionary terms from evolved traits whose distribution is not solely explained in these terms is not theoretically dubious because these traits call for different types of explanation.

## 1.7 Conclusion

An intense theoretical effort is ongoing to reconstruct the notion of human nature in light of progress in evolutionary biology and genetics; and the nomological notion of human nature is one of the candidate successors to the discredited essentialist notion of human nature. The scrutiny it has received in recent years has not undermined its appeal. The notion of a trait, which was not carefully specified in Machery (2008), should be understood broadly: traits—including dispositions—of typical groups of human beings, and of human beings at particular stages of a typical life history or that persist throughout, count as being part of human nature. The two components of the notion of human nature—the universality and the evolution proposals—have both been defended against criticism here. While typicality has no special status in biology influenced by population thinking, it is an explanandum that can be justifiably singled out from perspectives other than population thinking. The evolution proposal was clarified in this chapter: the proposal is not to distinguish two types of traits based on their ontogeny; rather, it is to distinguish those traits that can be proper targets of explanations appealing to evolutionary information from those that cannot.

## References

Amundson, R. (2005). *The Changing Role of the Embryo in Evolutionary Thought: Roots of Evo-Devo*. Cambridge: Cambridge University Press.

Antony, L. M. (1998). 'Human Nature and Its Role in Feminist Theory.' In J. A. Kourany (ed.), *Philosophy in a Feminist Voice: Critiques and Reconstructions*, 63–91. Princeton, NJ: Princeton University Press.

Boyer, P. (2001). *Religion Explained: The Evolutionary Origins of Religious Thought*. New York: Basic Books.

Buller, D. J. (2005). *Adapting Minds: Evolutionary Psychology and the Persistent Quest for Human Nature*. Cambridge, Mass.: MIT press.

Carnap, R. (1950). *The Foundations of Probability*. Chicago: University of Chicago Press.

Downes, S. (2017). 'Human Nature: An Overview.' In R. Joyce (ed.), *Routledge Handbook of Evolution and Philosophy*, 155–66. Abingdon: Routledge.

Duchaine, B., Cosmides, L., and Tooby, J. (2001). 'Evolutionary Psychology and the Brain.' *Current Opinion in Neurobiology* 11: 225–30.

Ghiselin, M. T. (1997). *Metaphysics and the Origins of Species*. Albany: State University of New York Press.

Gintis, H. (2008). 'Punishment and Cooperation.' *Science* 319: 1345–6.

Griffiths, P. E. (2009). 'Reconstructing Human Nature.' *Arts* 31: 30–57.

Griffiths, P. E. (2011). 'Our Plastic Nature.' In S. Gissis and E. Jablonka (*eds*), *Transformations of Lamarckism: From Subtle Fluids to Molecular Biology*, 319–30. Cambridge, Mass.: MIT Press.

Griffiths, P. E. and Machery, E. (2008). 'Innateness, Canalization, and "Biologicizing the Mind".' *Philosophical Psychology* 21: 397–414.

Griffiths, P., Machery, E., and Linquist, S. (2009). 'The Vernacular Concept of Innateness.' *Mind and Language* 24: 605–30.

Henrich, J., and Gil-White, F. J. (2001). 'The Evolution of Prestige: Freely Conferred Deference as a Mechanism for Enhancing the Benefits of Cultural Transmission.' *Evolution and Human Behavior* 22: 165–96.

Herrmann, B., Thöni, C., and Gächter, S. (2008). 'Antisocial Punishment across Societies.' *Science* 319: 1362–7.

Hill, R. A., and Dunbar, R. I. (2003). 'Social Network Size in Humans.' *Human Nature* 14: 53–72.

Hull, D. L. (1986). 'On Human Nature.' *Proceedings of the Biennial Meeting of the Philosophy of Science Association* 2: 3–13.

Kelly, D. (2011). *Yuck! The Nature and Moral Significance of Disgust*. Cambridge, Mass.: MIT Press.

Klasios, J. (2016). 'Evolutionizing Human Nature.' *New Ideas in Psychology* 40: 103–14.

Kronfeldner, M., Roughley, N., and Toepfer, G. (2014). 'Recent Work on Human Nature: Beyond Traditional Essences.' *Philosophy Compass* 9: 642–52.

Kurland, J. A., and Gaulin, S. J. (2005). 'Cooperation and Conflict among Kin.' In D. Buss (ed.), *The Handbook of Evolutionary Psychology*, 447–82. Hoboken, NJ: Wiley.

Laland, K. N., Sterelny, K., Odling-Smee, J., Hoppitt, W., and Uller, T. (2011). 'Cause and Effect in Biology Revisited: Is Mayr's Proximate-Ultimate Dichotomy Still Useful?' *Science* 334: 1512–16.

Lennox, J. G. (2001). *Aristotle's Philosophy of Biology: Studies in the Origins of Life Science*. Cambridge: Cambridge University Press.

Lewens, T. (2012). 'Human Nature: The Very Idea.' *Philosophy and Technology* 25: 459–74.

Lewens, T. (2015). *Cultural Evolution*. Oxford: Oxford University Press.

Lewis, C. J. (unpublished). 'Is the Nomological Account of Human Nature Committed to Essentialism?'

Linquist, S., Machery, E., Griffiths, P. E., and Stotz, K. (2011). 'Exploring the Folkbiological Conception of Human Nature.' *Philosophical Transactions of the Royal Society of London B* 366: 444–53.

Lipton, P. (1990). 'Contrastive Explanation.' *Royal Institute of Philosophy Supplement* 27: 247–66.

Liu, D., Wellman, H. M., Tardif, T., and Sabbagh, M. A. (2008). 'Theory of Mind Development in Chinese Children: A Meta-Analysis of False-Belief Understanding across Cultures and Languages.' *Developmental Psychology* 44: 523–31.

Machery, E. (2008). 'A Plea for Human Nature.' *Philosophical Psychology* 21: 321–30.

Machery, E. (2012). 'Reconceptualizing Human Nature: Response to Lewens.' *Philosophy and Technology* 25: 475–8.

Machery, E. (2016a). 'Human Nature.' In H. L. Miller (ed.), *The SAGE Encyclopedia of Theory in Psychology*, 433–5. Thousand Oaks, Calif.: Sage.

Machery, E. (2016b). 'Interview about Human Nature.' In A. Fuentes and A. Visala (ed.), *Conversations about Human Nature*, 55–68. Walnut Creek, Calif.: Left Coast.

Machery, E. (2016c). 'Human Nature.' In D. Livingstone Smith (ed.), *How Biology Shapes Philosophy: New Foundations for Naturalism*, 204–26. Cambridge: Cambridge University Press.

Machery, E. (2017). *Philosophy within Its Proper Bounds*. Oxford: Oxford University Press.

Machery, E. and Barrett, C. (2006). 'Debunking *Adapting Minds*.' *Philosophy of Science* 73: 232–46.

Mayr, E. (1961). 'Cause and Effect in Biology.' *Science* 134: 1501–6.

McElreath, R., Boyd, R., and Richerson, P. J. (2003). 'Shared Norms and the Evolution of Ethnic Markers.' *Current Anthropology* 44: 122–30.

Mithen, S. (2005). *The Singing Neanderthal*. London: Weidenfeld & Nicolson.

Odenbaugh, J. (unpublished). 'Human Nature, Anthropology, and the Problem of Variation.'

Olalde, I., Allentoft, M. E., Sánchez-Quinto, F., Santpere, G., Chiang, C. W., DeGiorgio, M., Prado-Martinez, J., Rodríguez, J. A., Rasmussen, S., Quilez, J. and Ramírez, O. (2014). 'Derived Immune and Ancestral Pigmentation Alleles in a 7,000-Year-Old Mesolithic European.' *Nature* 507: 225–8.

Powell, R. (2012). 'Human Nature and Respect for the Evolutionarily Given: A Comment on Lewens.' *Philosophy and Technology* 25: 485–93.

Ramsey, G. (2013). 'Human Nature in a Post-Essentialist World.' *Philosophy of Science* 80: 983–93.

Ramsey, G., and De Block, A. (2017). 'Is Cultural Fitness Hopelessly Confused?' *British Journal for the Philosophy of Science* 68: 305–28.

Richerson, P. J., and Boyd, R. (2005). *Not by Genes Alone: How Culture Transformed Human Evolution.* Chicago: University of Chicago Press.

Samuels, R. (2012). 'Science and Human Nature.' *Royal Institute of Philosophy Supplement* 70: 1–28.

Shyue, S. K., Hewett-Emmett, D., Sperling, H. G., Hunt, D. M., Bowmaker, J. K., Mollon, J. D., and Li, W. H. (1995). 'Adaptive Evolution of Color Vision Genes in Higher Primates.' *Science* 269: 1265–7.

Slater, M. (2013). *Are Species Real?* London: Palgrave Macmillan.

Sober, E. (1980). 'Evolution, Population Thinking, and Essentialism.' *Philosophy of Science* 47: 350–83.

Sperber, D. (1996). *Explaining Culture: A Naturalistic Approach.* Oxford: Blackwell.

Sperber, D., and Hirschfeld, L. A. (2004). 'The Cognitive Foundations of Cultural Stability and Diversity.' *Trends in Cognitive Sciences* 8: 40–46.

Stotz, K. (2010). 'Human Nature and Cognitive-Developmental Niche Construction.' *Phenomenology and the Cognitive Sciences* 9: 483–501.

Van Fraassen, B. C. (1980). *The Scientific Image.* Oxford: Oxford University Press.

Wachbroit, R. (1994). 'Normality as a Biological Concept.' *Philosophy of Science* 61: 579–91.

Wengrow, D. (2013). *The Origins of Monsters: Image and Cognition in the First Age of Mechanical Reproduction.* Princeton, NJ: Princeton University Press.

Winsor, M. P. (2003). 'Non-Essentialist Methods in Pre-Darwinian Taxonomy.' *Biology and Philosophy* 18: 387–400.

Winsor, M. P. (2006). 'The Creation of the Essentialism Story: An Exercise in Metahistory.' *History and Philosophy of the Life Sciences* 28: 149–74.

Wrangham, R. (2009). *Catching Fire: How Cooking Made Us Human.* Philadelphia: Basic Books.

# 2

# Trait Bin and Trait Cluster Accounts of Human Nature

*Grant Ramsey*

## 2.1 Introduction

Consider the mineral hematite, one of the chief sources of iron.* What is the nature of hematite? One answer might involve reciting its formula, $Fe_2O_3$. This formula points to one of hematite's invariant properties, its ratio of two iron atoms for every three oxygen atoms. But stopping at the molecular formula would be rather unsatisfying—it offers only modest help in distinguishing hematite from other minerals, especially ones that can appear similar, such as magnetite. Furthermore, because $Fe_2O_3$ can occur in various polymorphs such as maghemite—which has a cubic crystalline structure, unlike the rhombohedral structure associated with hematite—$Fe_2O_3$ does not uniquely identify hematite. Manifest properties, such as its colour or texture, can add valuable information, though in this case they can vary significantly. Hematite can appear dull or lustrous, and its colour can vary from black to grey to pale orange. The highly variable nature of these properties undercuts their use in helping to identify hematite. What are more stable are properties that would be realized only given certain counterfactual situations. For example, one invariant feature of hematite is the colour of its streak. The streak colour of a mineral is the colour achieved by scraping it on a streak plate, which is standardly a piece of unglazed porcelain. Thus, part of the nature of hematite might consist in what we would achieve were we to abrade it on a streak plate. Many other properties are of this kind. Were we to heat a sample of the mineral, at what point would it melt? Hematite nature is thus about the evident properties that samples of hematite bear; but also, and perhaps more

---

* Thank you to Michael Deem, Hugh Desmond, Tim Lewens, and Edouard Machery for taking the time to carefully read and comment on earlier drafts of this chapter. The chapter was completed while I was on a National Endowment for the Humanities-supported fellowship at the National Humanities Center. I thank the NEH and NHC for their support. Any views, findings, conclusions, or recommendations expressed in this chapter do not necessarily reflect those of the National Endowment for the Humanities.

importantly, it concerns properties that samples do not bear but possibly would bear if certain things were to happen to them.[1]

Let's label the possible things that can happen to hematite the 'possible hematite lives'—the possible life histories that pieces of hematite could live, being crushed, weathered, heated, ground, suspended in water, and so on. These possible lives will exhibit patterns from which generalizations can be drawn. If streaked, a red powder will appear. If heated to approximately 1,500 degrees Celsius, it will liquefy. We could thus imagine the set of possible hematite lives, and the properties of those lives. Lives that involve the property of being heated in excess of 1,500 degrees Celsius will result in the consequent trait of being a liquid; having a certain force applied will result in the sample being crushed; and so on. (These of course require certain 'normal' background conditions to obtain, conditions that are implicit here, but could be made explicit.) These generalizations are possible because of the patterns of traits that exist over the set of possible life histories. I suggest that we can understand hematite nature as consisting in the clusters of traits dispersed over hematite possible life histories. Hematite nature is not a bin of traits, but instead a set of relations among traits. Call this the 'trait cluster account' of hematite nature.

The main claim of this chapter is that the trait cluster account is not just a good account of hematite nature, or mineral nature broadly considered, but also serves as an adequate account of human nature. In the case of hematite, it may be possible to describe its nature in terms of simple dispositions, where these traits are (largely) unmanifested dispositions. But when one considers biological organisms and their complex life histories, such dispositions will not work to capture the complex ways in which traits are related to one another. Instead, a more nuanced understanding of trait patterns over life histories will be required. Rather than talking of a disposition to anger, say, it is more informative to take the trait of being angry and see which other traits it is related to and how they are related. My argument here for a life history-centred account of human nature is an expansion of my earlier (2013) proposal. In addition to articulating this account, I will draw out many of its implications and show how it enables conceptions of human nature to align with the human sciences. In defending the trait cluster framework, I will contrast it with the 'trait bin' approach, which characterizes human nature as a bin of traits instead of the relationships among traits.

Both the trait bin and trait cluster accounts are responses to the felt need for a concept of human nature that avoids the perils of essentialism. As Hull (1986) noted, since biological species (including *Homo sapiens*) are not defined in terms of essences, we cannot use species essences to construct a concept of human nature. The conclusion Hull drew was a sceptical one: he was sceptical about the project of producing and using a concept of human nature. The trait bin and trait cluster accounts attempt to avoid the perils of essentialism, while retaining the concept of human nature.

---

[1] For information on the properties of hematite, see http://geology.com/minerals/hematite.shtml.

## 2.2 The Trait Bin Account of Human Nature

There are two fundamentally distinct approaches to understanding the relationship between human traits and human nature. One approach—the trait bin approach—holds that human traits fall into two mutually exclusive bins: the nature category and the other category (containing, say, cultural or learned traits). The second approach—the trait cluster approach—takes human nature to consist not in a bin of traits, but in patterns of trait expression. The trait bin account will be considered here, and the trait cluster account will be discussed in the following section.

Essentialist views of human nature are trait bin accounts. They place in the nature bin the traits essential to our species. But one need not be an essentialist in order to defend the trait bin approach. There are criteria other than being considered essential that one can use to sort traits into the nature category. In this section I will consider and critique a recent attempt at defending a non-essentialist trait bin account, that of Machery (2008, 2016, Chapter 1 this volume). In so doing, I will not be suggesting that his conception is a poor rendering of the trait bin approach. Indeed, it is a clever attempt to save the trait bin approach from the problems of essentialism. Instead, my critique extends beyond Machery's approach, showing that the trait bin approach itself is where the problems lie.

Machery argues that we should retain the human nature bin, but define it in terms other than essences. His proposal is that 'human nature is the set of properties that humans tend to possess as a result of the evolution of their species' (2008: 323). Thus, traits fall into the human nature bin just in case they occur in '*most* humans' (2008: 323)—which I will understand to be more than 50 per cent (though the precise percentage is not important for my critique)—and are a product of the evolution of our species. By 'human', he means humans now, not ones that existed in our evolutionary past or ones that will exist in our future. And by 'the evolution of their species' he does not mean only the traits that arose in our species, that is, since the origin of our species' lineage around two hundred thousand years ago. Instead, he means the evolution of our species as well as of the species ancestral to ours. Thus, a trait like maternal care is, for Machery, in the human-nature bin even though it preceded our species and is universal among mammals. The reasoning behind these criteria is that if human nature is meant to characterize humans, the included traits should be common. And by excluding traits not part of our evolution, Machery excludes traits 'exclusively due to enculturation or to social learning' (2008: 326).[2] Human nature is thus the common traits resulting from our evolutionary heritage.

---

[2] I should note that Machery has modified his criteria in his contribution to this volume. Partly in response to my earlier critiques (Ramsey 2013), he has made his account closer to my life-history trait cluster account: 'it is natural to think that describing what human beings are like involves describing what they are like at the various life stages that constitute the life history of human beings' (Machery, Ch. 1 this volume, p. 23). And he also adds traits possessed not by individuals alone, but ones that 'typical human groups tend to possess because of the evolution of their species' (p. 23). These new additions, though improvements, do not undercut my critiques of the trait bin approach.

There are many difficulties with these criteria. Although the 50 per cent dividing line has intuitive appeal, many important human traits are possessed by a minority of humans. Women typically undergo menopause, and this appears to be a particularly interesting derived trait in our species. But because fewer than 50 per cent of humans will undergo menopause, it is not part of human nature on Machery's view. Applying this criterion to hematite, the orange streak it leaves on a streak plate will not be a part of hematite nature, since only a tiny fraction of hematite in the world has undergone or will undergo a streak. Machery could, of course, bite the bullet and argue that just because the streak is an important diagnostic tool does not mean that it is a part of hematite nature. But it is not clear that this is a bullet we need to or should bite.

The criterion that traits in the human nature bin must be a result of the evolution of our species and not 'exclusively due to enculturation or to social learning' also has intuitive appeal. But which traits are exclusively due to enculturation or to social learning? Machery uses the example of the belief that water is wet: 'the belief that water is wet is not part of human nature, in spite of being common, because this belief is not the result of some evolutionary processes. Rather, people learn that water is wet' (2008: 327). Is Machery right that there is nothing related to the evolution of our species that helps explain such a belief? This is of course a difficult empirical question, but it does not seem outlandish to think that evolved cognitive capacities shared by the majority of humans are involved in the formation of such beliefs. But if this is true, it is hard to think of beliefs (or other human traits) that are 'pure' and free of any evolutionary influence. While Machery (2008: 327) anticipates this difficulty in acknowledging that 'it is probably correct that evolutionary processes causally contribute to the existence of any trait that is common among humans', he does not answer the concern that there are deep problems inherent in trying to divide traits into two mutually exclusive categories, one due to evolution, the other not due to evolution.

In Chapter 1 of this volume, however, Machery addresses this issue. He argues that 'the nomological account requires that one be able to distinguish those traits whose distribution is explained by appealing only to individual or social learning from other traits, not traits whose development is only influenced by individual or social learning from other traits' (p. 31). This is a slippery move, since given the pragmatics of explanation, it is possible that in many cases we will be satisfied with an explanation of a trait distribution that only appeals to learning. It nevertheless should be acknowledged that explanations of the distribution of traits can often benefit by pointing to evolved psychological traits. For example, we might notice that in the English language the more common words tend to have fewer syllables than rarer words. We might be satisfied in explaining this distribution in word size by pointing out that this was true of the previous generation, and that the current generation learned their language for the most part from that generation. The distribution of words length is thus explained purely by learning. Though some may find this satisfying, I think that pointing to psychological features of our species (attention spans, abilities to remember long words, and so on) may be more insightful in explaining such a distribution.

Another question concerns the criteria we should use to classify trait tokens into trait types. To say of a human behavioural token that it is part of an evolutionary process and has been a target of selection requires that this trait token be subsumed into a class of tokens that can be identified across generations. Machery uses the example of shame as a trait with an evolutionary history. Shame, of course, is manifested by particular shame tokens like 'feeling shameful because of a failure to make the honour roll at school'. Is such a behaviour the result of an ultimate explanation? This depends on how one partitions traits. If the group of traits includes only ones involving shame about school-related performance, then this is unlikely to have a selective history. That is, there was probably not a sequence of ancestors that exhibited this particular behaviour and received a fitness boost because of it. Instead, if we subsume such behaviour into a broader class of traits that share a common aetiology, in this case behaviours that are strongly connected to the emotion shame, then we might be able to offer an evolutionary account. Returning to the example of knowing water to be wet, the questions for Machery are to which class of traits this belongs, what unifies this class, and what justifies grouping the tokens into this class and not another. Once these questions are answered, we can ask whether the disposition uniting the tokens has been selected for. One intuitively subsumes shame-laden behaviour into the shame-caused category, but what is the right category for other traits like knowing water is wet? Machery needs to offer a framework for connecting trait tokens (token instances of shame or beliefs about water) with underlying dispositions before making his conclusion about which ones are the result of an evolutionary history.

With Machery's criteria, it is also difficult to know how to categorize quantitative traits (traits that can take on a range of values) instead of qualitative traits (traits an individual either possesses or does not). Take human height. Is it human nature to have a specific height? Heights have changed considerably over the past century or so, in part due to cultural (especially dietary) changes. In the Netherlands, for example, male height has increased by almost 13 per cent since the mid-nineteenth century, a fairly dramatic increase (Schönbeck et al. 2013). A quantitative trait like height thus raises two problems for Machery. First, does the fact that each human has a unique height (measured at a very fine grain) mean that height is not a part of human nature? Second, does the fact that the environment (cultural or otherwise) plays an important role in the trait expression mean that it is not part of human nature? Since many important human traits are quantitative, it seems that they should not be eliminated from human nature. Such questions pose deep problems for a trait bin account of human nature. A better approach to answering these questions would be to investigate the range of values of human height, and the way these values are related to such things as diet. Such a way of understanding human nature is what will be argued for below. But first, let's take stock of the critique of Machery's position and what it implies for the tenability of a trait bin approach.

Machery's account was challenged on the basis of how it attempts to sort traits into the human nature bin. His approach has problems generated by including only majority traits, and by excluding traits that are clean of evolutionary influence. I suggest that it is not just Machery's rendering of the trait bin approach that is problematic, but the approach itself. To see this, let's step back and consider how traits come about.

All human phenotypic traits are the result of development. They represent a complex interplay between our genes and environment (including our cultural environment), both of which have been passed down from our ancestors. Some traits are relatively well buffered against a range of differences in genes; others consistently develop over a wide range of environments. The degree to which environmental and genetic differences affect any particular trait can take on a wide range of values. Furthermore, the degree to which differences in properties of organisms can be linked to genetic differences critically depends on the background environment. Some environments set up organisms to be very sensitive to particular genetic differences, while other environments may render these genetic differences impotent. My suggestion is that if we consider the set of possible gene–environment combinations and the traits that these generate, there is no justifiable, non-arbitrary way of sorting these traits into two bins, one bin of traits belonging to the nature of our species, the other bin of traits not so designated. Because of this, I hold that the trait bin approach to characterizing human nature should be abandoned.

Machery's response (Chapter 1) to this argument counters that the explanation of trait distributions is always contrastive and that some traits can only be explained in terms of social or individual learning: 'to explain why some traits are typical, one can appeal to evolutionary information; to explain why other traits are typical, one can only appeal to individual or social learning. The former, but not the latter, are part of human nature' (p. 34). Machery provides an example of something that can only be explained in terms of social learning:

For instance, if a few years ago most human beings had learned Psy's song 'Gangnam Style', one would not provide a proper explanation of this explanandum by appealing to the evolution of the larynx and the vocal chords or the evolution of the human musical sensitivity (e.g. Mithen 2005) since evolutionary information would not be specific enough to distinguish between between Psy's song 'Gangnam Style' and Carly Rae Jepsen's 'Call Me Maybe'. (p. 33)

But how, I wonder, does Machery rule out explanations based on evolved traits? How any particular song spreads is going to be a function of such variables as the characteristics of the song and the underlying neurophysiology. Perhaps 'Gangnam Style' resonated with people more than 'Call Me Maybe' because it stimulated our (evolved) limbic system in ways that gave us greater pleasure or excitement. Pointing this out may not be a very satisfying explanation, but it suggests that we cannot simply assume a priori that explaining the difference in song performance will never benefit from citing evolved characteristics. Similarly, why so many Americans know 'The Star-Spangled Banner'—another of Machery's examples—surely involves, at least in part, (evolved) traits concerning tendencies to conform with those around you. Thus, while I agree that we are sometimes satisfied with explanations which point out that something was socially or individually learned, many times we want to go further, to know *why* one thing was learned and not another.

In sum, the trait bin approach is subject to a host of problems. My solution for human nature—the trait cluster account—will be explored in the next section.

## 2.3 The Trait Cluster Account of Human Nature

In Hamlin Garland's autobiography, he notes that 'to rob me of my memories of the circus would leave me as poor as those to whom life was a drab and hopeless round of toil. It was our brief season of imaginative life. In one day—in a part of one day—we gained a thousand new conceptions of the world and of human nature' (Garland 1917: 137). Circuses are known for their idiosyncratic characters and odd, if not bizarre, behaviours. Thus, we are faced with a dilemma. Either Garland and others who think we can learn about human nature by studying the full range of human behaviour are wrong, or Machery's trait bin approach is wrong, and it is not the high frequency of traits across the whole species that is essential to human nature.

In this section, I articulate and defend what I call the trait cluster approach to human nature. This approach requires a gestalt shift in how human nature is understood. Traits are not taken to be within or outside of human nature. Instead, human nature lies within the patterns of expressions of the traits. To see this shift, consider how we would characterize the nature of a baseball. Under the trait bin approach—Machery's version of it at least, which he labels the *nomological* account—we would look for properties that are frequent among baseballs and a part of baseball history. Traits included in the baseball nature bin might be spending most of their time sitting still indoors or being made of leather, since these traits, though not universal, are a part of the majority of baseballs. By contrast, the trait cluster approach does not hold that there is a bin of traits constituting baseball nature. Instead, the nature of a baseball lies in how it performs in various circumstances. When hit with a bat with a particular angle and velocity, how much is it deformed and how far will it travel? Baseballs do not act outside of their nature, but this does not mean that they do not have a nature, or that they have an overly permissive nature. Baseballs act in very specific, reliable ways given their particular environmental inputs, and it is their nature to do so. Paralleling the hematite example above, nothing samples of hematite do goes against hematite nature, but this does not mean that hematite nature is overly permissive or vacuous. Instead, hematite trait clusters are very specific, and very informative in distinguishing hematite from other minerals and in explaining patterns of traits among samples of hematite.

Before articulating in detail my account of human nature, let's consider whether there have been other trait cluster conceptions of human nature. I will focus here on Griffiths (2009), Cashdan (2013), and Samuels (2012), since their views come the closest to my own by placing patterns of trait variation centre stage.

Griffiths argues that 'the primary sense which should be attached to the term "human nature" is simply what human beings are like, not some cause that makes them that way. As such, human nature is primarily the pattern of similarity and difference amongst human beings' (2009: 53). Griffiths is arguing against a hidden, inner nature that causes manifest traits. This argument is born out of his anti-essentialism: it is simply false that each human contains within himself or herself the essence of our species, and that this essence causes our behaviour. Griffiths, however, does not adopt Hull's (1986) sceptical solution that there is no such thing as human nature, nor does he accept Machery's solution of offering an anti-essentialist trait bin account. Instead, human nature is about *patterns* of similarity and difference among human beings. The account I will offer goes a bit further than this, taking human nature to reside not just in the patterns of similarity and difference across human beings, but also in the way traits are patterned within individual life histories. But before fleshing this idea out, let's consider another trait cluster account, that of Cashdan. As she argues,

because human nature evolved to be flexible in predictable ways, the task of understanding human nature requires that we understand how evolution shaped that variation. The assumption is not just that we evolved flexibly, but that selection shaped the nature and direction of that flexibility. To a behavioral ecologist, then, the predictable, patterned nature of that response is the universal we must understand. In this view, we cannot understand our universal human nature without understanding the variability in its expression […] The concept is clarified by viewing variation as a norm of reaction—the pattern of expression of a genotype across a range of environments.  (Cashdan 2013: 71)

Cashdan holds that human nature consists not in a set of traits, but instead in patterns of trait expression. She is thus rejecting a trait bin account and endorsing the trait cluster approach. She is primarily interested in one kind of pattern: how genotypes are expressed differently over a range of environments. Understanding human nature, then, is based on understanding how evolution shaped our ability to flexibly and predictably adapt to different environments. While I agree with Cashdan that patterns of gene expression across environments are an important component of the trait cluster account of human nature, they do not exhaust human nature.

One other account I should mention before introducing my version of the trait cluster account is that of Samuels (2012). He labels his view essentialist, but it is not essentialist in the way the term is standardly understood. As Samuels argues, 'human nature is a suite of mechanisms that underlie the manifestation of species-typical cognitive and behavioral regularities' (2012: 3). This view is in some ways not too far off from my own, since it concerns patterns of traits and the investigation of what underlies these patterns. But I don't think that it is helpful to restrict human nature to mechanisms. First, this intertwines human nature with debates about what a mechanism is. Second, as Machery (2016) points out, if we want human nature to describe what we are like, we should not restrict human nature to mechanisms. Of course, Machery holds that one should instead offer a bin of traits, whereas I hold that one should be concerned with how the traits are associated. I will now describe in detail my version of the trait cluster account.

I previously proposed a trait cluster account of human nature (Ramsey 2013), which I labelled the 'life-history trait cluster' (LTC) account because it is based on patterns of traits over sets of human life histories. I will retain the LTC label in order to differentiate my account from other trait cluster accounts like that of Griffiths and Cashdan. The idea behind the LTC account is this: each individual human lives his or her life, and in so doing realizes a life history. This life history is their path through space and time, and contains all the traits they have exhibited over their lifetime. The one life history that any particular human realizes is not the only way he or she could have lived over his or her life. Given the genes and the environmental arena in which the individual resides, there is a set of possible ways that he or she could have lived his or her life. (This is not a claim about indeterminism—instead one can speak of possibilities in deterministic systems, like possible games of pinball given possible starting states and game plays.) Each of these possible life histories can be understood as a four-dimensional structure, and this structure is populated with traits. These organismic traits can occupy the whole life history, but will frequently occupy only a subset of each life history. Having blood coursing through our veins is something each of us will have for our entire lives, unless we experience a cardiac arrest from which we are able to recover. And being asleep is a trait that we experience during roughly one-third of our life history, cycling on and off. Other traits, like adult teeth, are not inborn (though their buds are), but once we have them they usually stick around for the entire remainder of the life history. 'Trait' in this sense is thus used quite liberally; even a hiccup, sneeze, or yawn issued at a particular time is a trait.

Let's now examine how this set of life histories links up with human nature. It is clear that for any individual, while they have—at a fine grain of analysis—an immense number of possible lives that they can lead, the pattern of traits is nevertheless rather constrained (holding fixed their genes and the extant environmental arena[3]). They are constrained in the obvious sense that some traits simply cannot arise—no human will hibernate or sprout wings and fly—but there are subtler and more interesting constraints as well. Traits that are a part of at least some possible life histories ('possible lives' for short) will nevertheless be limited by such things as age and the sequence of preceding traits. Adult teeth erupt following (and not before) baby teeth; mothers lactate following (and not before) pregnancy; pregnancy only happens between menarche and menopause. While there may be exceptions to such generalizations, there are nevertheless patterns to how the traits are distributed over life histories.

It is these patterns, and the generalizations we can derive from them, that constitute individual nature—the nature of an individual organism, that is. Similarly, the nature of a particular baseball is based on the possible ways it will behave over its life. If hit in certain sorts of ways, it will be propelled for long distances. If it is lost in the woods, it will slowly decompose, but will do so in a predictable way. If it is exposed to prolonged hot, dry

---

[3] If one does not hold these fixed, then it is no longer clear that it is human nature that is being articulated. More on this below.

weather, it will predictably form small cracks. Thus, we could map out the possible fates of the baseball, and note the pattern of trait expression over the set of possibilities. Such patterns, and the generalizations we can draw from them, constitute the nature of the ball. If the ball is hit but does not burst into flames, we can note that such a flight is not a part of this ball's nature. The baseball's nature can thus serve to explain its outcomes.

Because of the laws of physics, especially under some interpretations of quantum mechanics, the ball has a non-zero probability of (for example) spontaneously jumping from the pitcher's hand to the top of her head. This behaviour is within the ball's nature, though it has such a small probability that it can be ignored. Similarly, the nature of an individual human will include countless highly improbable events. And, as with the ball, they can safely be ignored despite their inclusion within the individual's nature.

Let's now define individual nature precisely. An individual's nature is the pattern of its trait expression over its set of possible life histories. (Again, this is holding fixed the individual's set of genes and environmental arena, and of course the laws of nature. If genes were allowed to vary, individual nature would be vacuous since sufficient changes to genes could, say, change an American into an aardvark. By contrast, varying the way that an individual encounters its environmental heterogeneity reveals something about its nature.) Individual nature thus includes both the patterns within individual life histories and the patterns across those life histories. Individual nature, however, is not human nature; it is instead its foundation. Consider the individual nature of each human. If we were to combine the possible life histories from each individual's nature, then we would have human nature as the patterns of trait expression over the totality of extant human possible life histories. Because humans are an evolving species—not something with a static nature like a chemical element—human nature is restricted to extant humans (in the sense of extant genomes and sets of environmental variables). The benefit of this is that it allows human nature to change over time, and does not require of human nature that it take into account the deep past.

By focusing on patterns of trait expressions, this account of human nature—the LTC account—does not directly concern the actual frequency of traits. While the frequency of traits can serve as evidence for human nature, actual frequency per se is not definitionally linked to human nature, as it is in some trait bin accounts. I argued in the introduction that the natures of scientific kinds like hematite are best defined not in terms of the bin of traits that most pieces of hematite possess, but instead in terms of the patterns of expressions of the traits within the space of possible hematite lives. These patterns carry more information about the nature of hematite, and are more helpful in distinguishing hematite from other, similar minerals. Hematite nature, and mineral nature broadly considered, are thus the subject of the mineralogical sciences. Scientists don't just passively observe and tally the manifest properties of minerals; instead, they do things to the minerals and see how traits like 'being bathed in hydrochloric acid' are associated with changes in the mineral. And just as an LTC account of mineral nature allows mineral nature to be aligned with the mineralogical sciences, so does the LTC account of human nature allow for human nature to be aligned with the human sciences. But before arguing for a link between the LTC account of human nature and the human sciences, let's step back and consider what roles human nature might be expected to fulfil, and whether the LTC account can fulfil them better than a trait bin account.

## 2.4 What is Human Nature Good For?

I have argued that there is a concept of human nature, the LTC account, that does not fall prey to the same difficulties as trait bin accounts. But in order to see whether the trait cluster approach is truly superior to the alternatives, we must pause to consider what the concept of human nature is for, what use it can have.

The fact that 'human nature' has currency in the academic and popular media urges us to respond in one of two ways. One way is to hold that 'human nature' is like 'phlogiston'—it is a term without a referent within the contemporary worldview. It should thus be discarded and replaced with one or more new concepts, just as phlogiston is no longer taken to be a substance, but instead the absence of oxygen.[4] Could human nature have a similar replacement? The problem with finding a replacement is that human nature does not play a specific, technical role in the sciences. It is less like a theoretical posit and more like a name for the general subject of the human sciences: human nature is the quilt and the various sciences concern patches on this quilt. If this is true, then we should pursue the second way of responding, which is to produce a clearer notion of what human nature is. Such a conception should (1) articulate the subject of the human sciences. In so doing, it might help to (2) characterize what human beings are like, and may even be used to (3) causally explain what we are like. Other desiderata might include (4) articulating human limits, (5) distinguishing humans from other animals, and (6) providing normative insight about how humans should be. In what follows in this section, I will consider desiderata (1)–(3) to see which of these roles the LTC can play, and compare the results to the roles played by Machery's nomological account. For a discussion of (4) and (5), see Ramsey (2013); for a discussion of (6), see Ramsey (2012, 2017).

### 2.4.1 Human nature as the subject of the human sciences

If human nature is not based on intuition or religious texts, but is instead based on what we do and how and why we do it, human nature should be aligned with the human sciences. We can thus consider human nature to be the subject of the human sciences. To see whether the LTC account of human nature is tenable—and superior to trait bin approaches—we should consider what the human sciences are after, and what kinds of results they consider to be worthy of reporting.

---

[4] This is to gloss over the complexities of the case of phlogiston—see Chang (2009).

The human sciences are diverse, and include such disciplines as psychology, sociology, anthropology, and economics. In comparing trait bin and trait cluster accounts, the task at hand is to see whether these sciences are primarily concerned with how traits are related to one another, and whether the sciences restrict their domain of enquiry to (in Machery's rendering) traits that are possessed by the majority of humans and are due to human evolution. Let's focus on one of the sciences, psychology. Machery's (2008) paper was published in the journal *Philosophical Psychology*, and he certainly has psychology as a target science. Should we then conclude that psychological science centres on traits that are evolutionary products possessed by the majority of humans? This is an empirical question, and the rigorous survey of the psychology literature capable of decisively answering it is beyond the scope of this chapter. But to get a flavour of what psychology tends to focus on, let's consider the subject matters of the papers published by the official journal of the American Psychological Society, *American Psychologist*, and whether they tend to concern the bin of traits identified by Machery as human nature or instead associations among traits.

For a sample, let's examine the first article in each issue of volume 70 of *American Psychologist* (up to October 2015, which is the most recent volume at the time of writing). The first article of issue 1 concerns the promises of qualitative inquiry, so it does not directly bear on the matter of human nature. The second article, 'Evaluating Gender Similarities and Differences Using Metasynthesis', concerns ways of evaluating how similar traits are across genders. This is clearly about trait clusters, and is not about the majority of traits that are products of our evolution. Moving on to issue 2, the first article is 'Cancer Control Falls Squarely within the Province of the Psychological Sciences' (Green McDonald et al. 2015). This article concerns topics like 'evidence linking certain behaviors to cancer risk and outcomes' (2015: 61). The article is thus clearly about associations among particular behavioural traits and certain forms of cancer occurring in the minority of the population. The first article in issue 3 is 'National Accounts of Subjective Well-Being', which discusses associations among well-being traits and policy-relevant factors. Issue 4, a special issue on bullying, clearly concerns trait associations. For example, the first article after the introduction, 'Long-Term Adult Outcomes of Peer Victimisation in Childhood and Adolescence: Pathways to Adjustment and Maladjustment' (McDougall and Vaillancourt 2015), is about how childhood bullying traits are clustered with traits in adults. Issue 5 concerns lists of achievements, awards, and obituaries. The first article of issue 6 is about the replication crisis and the role of sample size, and is thus less directly relevant to the issue at hand. Issue 7 is a special issue on mindfulness. The first article, 'Conceptual and Methodological Issues in Research on Mindfulness and Meditation' (Davidson and Kaszniak 2015), is a methodological article that deals with issues such as 'the nature of control and comparison conditions for research that includes mindfulness or other meditation-based interventions' (2015: 581). This paper focuses on patterns of associations among traits, in this case between meditation-based interventions and other traits.

While it may be that this small survey of *American Psychologist* is not representative of the whole of psychology, or of the human sciences broadly considered, I consistently find in this and other samples of the literature studies that concern one kind of trait (meditating, say) and its association with other traits (like stress levels). I tend not to find studies that restrict their domain of inquiry to traits that the majority of humans possess, or traits that have an evolutionary heritage. Generalizing from this sample, I am confident that if we want human nature to be the subject of the human sciences, human nature must concern associations among traits, not a bin of traits. The LTC account, therefore, fulfils the desideratum of having human nature be the subject of the human sciences. Downes argues of trait cluster approaches that 'their most fruitful contributions are not their alternate characterizations of human nature. Rather, they each provide alternate, evolutionarily influenced frameworks for understanding and explaining human variation, both of which are valuable resources for social scientists confronting human variation' (2016: 919). But if human nature is the subject of the human sciences, it appears that the LTC account is a good approach to understanding human nature, and is certainly better from methodological and empirical standpoints than alternative trait bin accounts.

Machery and other trait bin theorists might reply that this is simply not a desideratum of a conception of human nature. Of the five traditional desiderata Machery (2016) lists, he does not include human nature being the subject of the human sciences. But if human nature is disconnected from the human sciences, two problematic conclusions follow. One is that human nature will have arbitrary boundaries. The 50 per cent criterion of Machery seems arbitrary, and using the sciences as our guide, we don't need to produce a priori boundaries like this. The second conclusion is that if human nature is free-floating and independent of the human sciences, then we have the odd situation that most human scientists are not concerned with discovering human nature; rather, they only accidentally discover features of human nature when they happen to focus on the right bin of traits. Furthermore, if the sample of papers from *American Psychologist* is indicative of the human sciences, scientists very rarely study human nature in Machery's nomological sense. Thus, under the nomological account, human nature and the subject of the human sciences are minimally overlapping. But if this is true, it is difficult to see what the point is of retaining the concept of human nature. In order to see why we should retain the concept of human nature, let's consider other desiderata and whether the trait bin approach might better satisfy them.

### 2.4.2 What are humans like?

Even if the conception of human nature that emerges from the nomological approach is not the subject of the human sciences, perhaps the approach is nonetheless useful for describing what we are like. Machery certainly thinks his account fulfils this desideratum: the nomological account 'fulfills the descriptive function; indeed, it was developed to fulfill it' (2016: 214). But does his approach really do an adequate job characterizing human nature in this purely descriptive way? One difficulty, mentioned above, is that many traits of central importance to our species do not occur in the majority of individuals. This is true of many species. It is an important feature of chimpanzees that there is a status hierarchy and that there are alpha males and alpha females. This does not lose its importance if we realize that only a minority of chimpanzees will ever achieve alpha status. Another difficulty is that so much of what humans are like involves, not a set of traits, but (as we've seen) the relation among traits. Just as describing a chemical compound by listing a bin of constituent elements will provide much less insight than noting how the elements are bonded to one another, so noting how human traits are bonded to one another and distributed over life histories is where the deep insights about human nature come from.

Nevertheless, Machery holds that his nomological approach is a superior account in characterizing humans:

> Because the nomological notion of human nature fulfills the descriptive function, human nature has predictive power: This allows scientists and lay people to make probabilistic predictions about how people are going to behave in particular situations. This is [in] line with the use of the notion of human nature in the sciences. For instance, Gintis [...] makes the following prediction about a behavioral-economics game: 'Because the four subjects are strangers, the standard view of human nature suggests that there will be zero contributions.' (2016: 215)

But is human nature as Gintis is using it a good fit with the nomological approach? Gintis's paper quoted by Machery concerns economic games studying punishment and cooperation across cultures. Gintis claims that 'the standard view holds that human nature has a private side in which we interact morally with a small circle of intimates and a public side in which we behave as selfish maximizers' (2008: 1346). Gintis then goes on (in the text quoted above by Machery) to note what this standard view entails about how strangers should interact. He later points out that 'in the many times this game has been played in a variety of social settings, the older view [of human nature] is virtually never supported' (2008: 1347). In the particular study discussed by Gintis, he points out that 'antisocial punishment was rare in the most democratic societies and very common otherwise' (p. 1346).

In sum, Gintis offers a simple conception of human nature and notes that it is not born out by the data. Instead, the data show that human nature is much more complex—that depending on the cultural milieu, different traits will predictably appear. Exposure to a democratic worldview results in one set of cooperation and punishment strategies, while an absence of democratic exposure results in another. Human nature in this case is best understood in terms of a pattern of economic-decision outcomes and their relation to prior life history traits. Such a finding about human nature is easily accommodated by the LTC account, but I fail to see how the nomological account has predictive power in this case. First, it is not clear that any one of these cooperation and punishment strategies exists in more than 50 per cent of the world population. Second, Machery's separation of human nature from culture and learning is problematic.

The results discussed by Gintis are clearly modulated by cultural influence and social learning, yet the predictable associations of upbringings and economic decision outcomes help to reveal human nature.

Machery uses the following case both to highlight the predictive power of his account, but also to criticize my trait cluster account:

> Ramsey's notion of human nature seems to have little predictive power. Because every phenotype that a human being could have belongs to one of the life histories included within human nature, on this notion one cannot justifiably infer that a human being is likely to possess a trait from the fact that this trait belongs to a life history included within human nature. (2016: 216)

Machery's critique misses the point. When we make predictions about which traits an individual will bear, we usually do not do so in the total absence of knowledge about them. Instead, we make predictions based on what other life history traits they possess. Following Gintis's discussion, we would predict one form of behaviour from individuals in democratic societies, another from individuals in other societies. In fact, one of the major conclusions of Gintis's discussion is that we need such information in order to make robust predictions about human behaviour. Thus, the fact that human life histories are heterogeneous is not a problem so long as we recognize that the heterogeneity is not random; it follows predictable regularities and results in predictable patterns of outcomes.

### 2.4.3 Causally explaining human characteristics

At the heart of the essentialist view is the idea that each of us has within us a uniquely human essence. And in at least some renderings of the essentialist account, we can cite this essence in explaining the nature, frequency, and distribution of human traits. But do non-essentialist trait bin accounts or the LTC account allow for causal explanations of human characteristics? As Machery points out, his conception of human nature 'is not viewed as a cause; rather, it is constituted by the outcomes of various evolutionary processes […] Thus, it is unclear how the notion of human nature could underwrite causal explanation of human beings' characteristics' (2016: 218).

Machery's account cannot causally explain outcomes, but he nevertheless holds that 'human nature can be a causal-explanatory notion despite not being a cause' (2016: 219), because human nature

> is an etiological kind: All the properties of human beings that are included in human nature have the same etiology in that they are the outcomes of evolutionary processes. As is the case with other etiological kinds, classifying a trait as belonging to human nature is thus to endorse a particular explanatory sketch: It is to assert that this trait is a proper target of an ultimate explanation. (2016: 220)

In other words, because his trait bin is composed only of traits that have an evolutionary history, to claim that a trait is part of human nature is to claim that it has an evolutionary history. Thus, any causal generalizations we can make about traits with evolutionary histories are generalizations that we can apply to human nature traits.

The causal-explanatory power of the nomological account is rather weak, but Machery thinks it does a better job than the LTC account: 'The causal-explanatory function is largely left unfulfilled by Ramsey's life-history trait cluster account of human nature. Every possible trait belongs to some life history included within human nature, and so asserting that a given trait is due to human nature provides no information at all. In this respect at least, human nature is not explanatory' (2016: 221). But Machery seems to be treating the LTC account as though it were a trait bin account. And he is right that if it were a trait bin account that included all traits, saying of a trait that it is in the human nature bin would carry no information. But the LTC account is a trait cluster account, not a trait bin account. There are therefore no traits within or outside of the human nature bin, since there is no such bin. Human nature consists in the relations among traits, not in features of a bin of traits.

Given that the LTC account is a trait cluster account and not a trait bin account, we can ask whether the LTC account, properly understood, can be used to causally explain human characteristics. If human nature is the subject of the human sciences, then the generalizations about trait clusters that are unearthed by science can serve as part of the explanans in explanations of human characteristics. And to the degree that such associations are causal, explanations in terms of these associations will be causal explanations. For example, if meditating lowers your stress, you can cite human nature to account for it, since psychologists have confirmed that meditation is associated with lower stress. And researchers do not stop at mere associations, but try to uncover reasons why the associations obtain; which causal pathways connect meditation behaviour with lower levels of stress? Thus, *pace* Machery, the LTC account not only can causally explain human characteristics, it also does a much better job than his nomological account.

## 2.5 Conclusions

Accounts of human nature fall into two broad categories, trait bin and trait cluster accounts. Trait bin accounts consider human nature to consist of a bin of traits. Trait bin accounts can be essentialist, in which the human nature bin is furnished only with traits essential to our being human. And there are non-essentialist trait bin accounts of human nature, like that of Machery (2008, 2016, Chapter 1 this volume), which define human nature as a bin based on properties other than essences.

The essentialist trait bin approaches are problematic because of the problems associated with attempting to maintain species essentialism in light of the non-essentialist contemporary phylogenetic classification system. Hull (1986) argued that the fact that natures are essences and that species are not defined in terms of essences shows that there is no such thing as human nature. Others have responded to this by arguing that there are non-essentialist trait bin or trait cluster conceptions of human nature that serve useful roles, and thus that the concept of human nature should not be abandoned.

The non-essentialist trait bin approach that I considered in detail here—Machery's nomological account—has merit, but bears a number of problems. The nomological account requires that for a trait to be included in the human nature bin, it must be exhibited by at least 50 per cent of the members of our species. This seems rather arbitrary; but more importantly, it leaves out many traits that are centrally important to characterizing our species and distinguishing it from others. And the requirement that the traits have an evolutionary, rather than cultural, cause runs into several difficulties. But beyond these difficulties, it is not clear that the nomological account can fulfil core desiderata for human nature, such as aligning human nature with the human sciences.

The failure of both essentialist and non-essentialist trait bin accounts motivates my trait cluster account of human nature, the LTC account. Trait cluster accounts hold that human nature lies not in which traits individual humans happen to have, but in the way the traits are exhibited over human life histories. In particular, traits are distributed in specific patterns over human life histories, and it is in these patterns where our nature lies. These patterns can be used to characterize humans, and the patterns of trait associations are precisely what the human sciences are concerned to uncover and explain. Thus, if we want to retain a conception of human nature that is capable of being the subject of the human sciences, it is the LTC account that we should adopt.

## References

Cashdan, E. (2013). 'What Is a Human Universal? Human Behavioral Ecology and Human Nature.' In S. M. Downes and E. Machery (eds), *Arguing about Human Nature: Contemporary Debates*, 71–80. New York: Routledge.

Chang, H. (2009). 'We Have Never Been Whiggish (about Phlogiston).' *Centaurus* 51: 239–64.

Davidson, R. J., and Kasznaik, A. W. (2015). 'Conceptual and Methodological Issues in Research on Mindfulness and Meditation.' *American Psychologist* 70: 581–92.

Downes, S. (2016). 'Confronting Variation in the Social and Behavioral Sciences.' *Philosophy of Science* 83: 909–20.

Garland, H. (1917). *A Son of the Middle Border*. New York: Macmillan.

Gintis, H. (2008). 'Punishment and Cooperation.' *Science* 319: 1345–6.

Green McDonald, P., O'Connell, M., and Suls, J. (2015). 'Cancer Control Falls Squarely within the Province of the Psychological Sciences.' *American Psychologist* 70: 61–74.

Griffiths, P. E. (2009). 'Reconstructing Human Nature.' *Arts* 31: 30–57.

Hull, D. L. (1986). 'On Human Nature.' *Proceedings of the Biennial Meeting of the Philosophy of Science Association* 2: 3–13.

Machery, E. (2008). 'A Plea for Human Nature.' *Philosophical Psychology* 21: 321–9.

Machery, E. (2016). 'Human Nature.' In D. Livingstone Smith (ed.), *How Biology Shapes Philosophy*, 204–26. Cambridge: Cambridge University Press.

McDougall, P., and Vaillancourt, T. (2015). 'Long-Term Adult Outcomes of Peer Victimization in Childhood and Adolescence: Pathways to Adjustment and Maladjustment.' *American Psychologist* 70: 300.

Mithen, S. (2005). *The Singing Neanderthals: The Origins of Music, Language, Mind, and Body.* London: Weidenfeld & Nicolson.

Ramsey, G. (2012). 'How Human Nature Can Inform Human Enhancement: A Commentary on Tim Lewens's "Human Nature: The Very Idea".' *Philosophy and Technology* 25: 479–83.

Ramsey, G. (2013). 'Human Nature in a Post-Essentialist World.' *Philosophy of Science* 80: 983–93.

Ramsey, G. (2017). 'What Is Human Nature For?' In A. Fuentes and A. Visala (eds), *Verbs, Bones, and Brains: Interdisciplinary Perspectives on Human Nature*, 217–30. Notre Dame, Ind.: University of Notre Dame Press.

Samuels, R. (2012). 'Science and Human Nature.' *Royal Institute of Philosophy Supplement* 70: 1–28.

Schönbeck, Y., Talma, H., van Dommelen, P., Bakker, B., Buitendijk, S. E., HiraSing, R. A., and van Buuren, S. (2013). 'The World's Tallest Nation Has Stopped Growing Taller: The Height of Dutch Children from 1955 to 2009.' *Pediatric Research* 73: 371–7.

# 3

# A Developmental Systems Account of Human Nature

*Karola Stotz and Paul Griffiths*

## 3.1 Current State of the Debate

The characteristics and causes of human nature constitute one of the oldest and most contested topics of inquiry. A scientifically credible account of human nature must assimilate and integrate findings from the biological, psychological, and social sciences. Contemporary philosophical work on human nature sets out to do this, but it also tries to stay in touch with older ideas about human nature. Almost all authors have accepted the Darwinian challenge and recognized that the human species is not defined by a fixed, inner essence. But despite this rejection of essentialism, many authors remain attached to the idea that human nature is confined to the left-hand side of the dichotomies between nature and nurture, innate and acquired, biology and culture (Machery 2008; Kronfeldner Chapter 10 this volume).[1]

This attachment reflects the fact that enquiries into human nature start from an everyday ('vernacular') idea of human nature and try to honour some of the intuitions associated with that idea. The vernacular conception of human nature is an expression of an implicit 'folk theory' of biological development, which has at its heart a distinction between traits that come from 'inside' and those imposed from 'outside'. We and our collaborators have conducted empirical research to characterize this folk theory in more detail (Griffiths 2002; Griffiths et al. 2009; Linquist et al. 2011). The folk theory of animal natures is an instance of 'psychological essentialism' (Medin and Ortony 1989; see also Gelman 2003) and the essential, inner nature of an animal is associated with traits that are fixed in development, typical of the species, and teleological—the animal is intended to have this trait. When this folk theory of animal natures is applied to humans, it produces the vernacular idea of human nature. We describe our 'three-factor' model and related psychological research in section 3.2.

---

[1] But see Downes and Machery (2013) for a collection of different views; Fuentes et al. (2010) for a collection of essays providing an anthropological challenge to a unitary theory of the human; and Lewens (2012a) for an extremely permissive, if not eliminativist, notion of human nature.

The problem with the vernacular idea of human nature is that it confounds three important but essentially independent biological properties. A trait can be fixed without being typical or having a purpose, it can be typical without having a purpose or being fixed, and it can have a purpose without being fixed or typical. This is one reason why so many developmental biologists and psychologists have rejected a simple dichotomy between innate and acquired characteristics (Lehrman 1953; Hinde 1968; Gottlieb 1970; Bateson 1991). The shortcomings of the vernacular idea of human nature are similar to the shortcomings of the pre-scientific concept of heat. Whether an object feels 'hot' depends on three physical quantities that can vary independently of one another—temperature, quantity of heat, and conductivity. Using these three, more precise ideas, we can explain what people are responding to when they say something is hot; but the original idea is not a useful construct with which to do science.

In this chapter, we defend a view of human nature that goes beyond the vernacular idea, in the same way that the physics of heat went beyond the phenomenological notion of things being hot. We argue that such an idea must fulfil several desiderata: it must be explanatory and not merely descriptive; it should make human nature an object of inquiry in the human sciences (all those disciplines that take the human species or some aspect of it as their subject, from physiology through psychology and anthropology to sociology); a science of human nature should explain the folk-biological features traditionally aligned with the idea of human nature in a way that makes clear why they won't do as defining features of human nature; and, lastly, our concept of human nature should embrace human diversity, plasticity, and polymorphism, because these are important aspects of the evolutionary design of human beings. We outline these desiderata in more detail in section 3.3.

We will argue that there are two extant theories that meet these requirements: Grant Ramsey's life history trait cluster (LTC) account (Ramsey 2013) and the developmental systems (DS) account of human nature (Griffiths 2011). In section 3.4, we outline the basic similarity between these two, namely, that both are grounded in human developmental biology. Both accounts suggest that to understand human nature is to understand the plastic but not unstructured process of human development.

While we are in agreement with much of Ramsey's account, in section 3.5 we draw attention to some differences between the two accounts. One major difference is that our account focuses more strongly on the human developmental environment as a critical factor in human nature. Drawing on the framework of developmental systems theory and the idea of developmental niche construction, we argue that human nature is not embodied in one input to development, such as the genome. The patterns of similarity and difference amongst human beings are explained by a human developmental system that reaches well out into the 'environment'.

We also emphasize that developmental systems theory creates a dynamical, process perspective on human nature. Human nature is underpinned by a range of mechanisms of extended inheritance, as well as genetic inheritance, and the life course of any individual human being depends upon a matrix of exogenetic developmental factors—the developmental niche. The fundamental unit of analysis in our approach is a process—a human life history (Griffiths and Stotz 2018).

## 3.2 The Folk-Biological Idea of Human Nature

Our account of the folk-biological conception of human nature builds on work in cognitive anthropology and child psychology that identified a pattern of essentialistic thinking—psychological essentialism—about living things across many human cultures and in human children (Atran 1990; Berlin 1992; Medin and Atran 1999; Medin and Atran 2004; Gelman 2003). It gains additional support from psychological research on the 'genetic essentialism framework' by psychologist Ilan Dar-Nimrod and collaborators. Our earlier work with our collaborators constructed a 'three-factor' model of folk-biological thought about animal natures; provided some experimental evidence for this model; and showed that in contemporary English, the idea of 'nature' is expressed by saying things are 'in the DNA' (Griffiths et al. 2009; Linquist et al. 2011). At around the same time, Dar-Nimrod and collaborators set out to study lay understandings of genetic causation, and documented a set of 'genetic essentialist biases' that correspond closely to elements of the three-factor theory of animal natures (Dar-Nimrod and Heine 2011a, 2011b; Dar-Nimrod and Lisandrelli 2012; Dar-Nimrod et al. 2012; Dar-Nimrod et al. 2014; Cheung et al. 2014).

The three-factor model proposes that there is a folk-biological, implicit theory of development in which some but not all characteristics of animals are expressions of a 'nature' inherited from their parents and which makes them the kind of animal that they are—a human, a chimp, or a kangaroo. Phenotypes that stem from this inner nature are expected to have three characteristics: fixity, typicality, and teleology. Fixity means that the phenotype is hard to change by environmental means. Typicality means that the phenotype is found in all or most members of the species (or of some natural subset such as a sex or an age group). Teleology means that the phenotype is part of the design of the organism. It is there for a reason, and organisms that lack these features are not how they are meant to be (see section 3.3, point 3 for a naturalistic interpretation of teleology). In Table 3.1 we show how these factors line up with elements of the genetic essentialist framework (GEF).

The GEF suggests that genetic attributions for various traits, conditions, or diseases activate four specific psychological processes, or genetic essentialist biases. The first bias, immutability/determinism, is that thinking about genetic attributions leads people to view relevant outcomes as less changeable and predetermined. To the extent that a phenomenon is perceived to be immutable, it will be perceived to be beyond someone's control. Genetic attributions decrease perceptions of control over relevant outcomes (Dar-Nimrod et al. 2012; Parrott and Smith 2014) and limit the perceived capability of other means, such as environmental manipulations or individuals' volition, to modify the outcome (Jayaratne et al. 2009). The second genetic essentialist bias, termed 'specific etiology', is a tendency to discount additional causal explanations once genetic

Table 3.1 Comparison between the genetic essentialism framework (GEF) and the three-factor model

| Genetic essentialist elements | Three-factor model of animal natures |
|---|---|
| (Dar-Nimrod and Heine 2011a) | (Linquist et al. 2011) |
| *Immutable and determined*: thinking about genetic attributions leads people to view relevant phenotypes as less changeable and predetermined | *Fixity*: phenotypes that are part of an animal's nature do not depend on the particular environment in which the organism is raised and are hard to change by environmental manipulations |
| *Specific etiology*: the tendency to discount additional causal explanations once genetic attributions are made | Traits are *either* expression of the animal's nature (and are expected to have the three features) *or* imposed by the environment (with opposite expectations) |
| *Homogeneous and discrete*: leads to a focus on the central identifying features that are common to all group members, drawing attention away from in-group differentiating features | *Typicality*: phenotypes that are part of an animal's nature are typical of the entire species or of some natural subset such as males or juveniles |
| *Nature*: phenotypes are perceived as a natural outcome (with positive normative associations) | *Teleology*: phenotypes that are part of an animal's nature serve some purpose (with positive normative associations) |

attributions are made. Hence, genetic attributions also increase the likelihood that people will disregard alternative casual attributions for complex phenomena (Dar-Nimrod and Heine 2011a). Whereas the first two genetic essentialist biases focus on individuals, the third, termed 'homogeneity/discreteness', concerns groups. Essentialist thinking leads people to focus on the central identifying features that are common to all group members, drawing attention away from in-group differentiating features. This leads people to view individual members of a category as more homogeneous, as they share the identifying features, which may contribute to stereotyping and more prejudiced attitudes toward group members (Dar-Nimrod and Heine 2011a). The final genetic essentialist bias is termed 'naturalness', i.e. genetic attributions increase the likelihood that a relevant outcome is perceived as a natural outcome. It is widely agreed in both philosophy and psychology that viewing an outcome as natural has important normative overtones.

The vernacular idea of human nature from which so many philosophical analyses start is simply the application of this form of essentialist thinking to humans. It seeks to divide human characteristics into those imposed by the environment and those that stem from an inner nature, and embodies the assumption that the three characteristics of fixity, typicality, and teleology are strongly associated with one another because traits that stem from our inner nature have these three properties and traits imposed by the environment do not.

However, this intuitive picture of biological development is fundamentally mistaken. All phenotypes are produced by a combination of genetic and environmental factors, and in many cases epigenetic factors. The patterns of interaction between these factors are many and varied, and do not conform to two distinct patterns, one of which is characteristic of traits that have been designed by natural selection. Some philosophers have conceded this, but suggested that there is a continuum, with evolved traits clustered at one end. However, in our view the plausibility of this idea comes not from reviewing the evidence, but from the continued influence of the folk-biological picture (Griffiths and Machery 2008; see also Mameli and Bateson 2006, 2011).

## 3.3 Desiderata for an Account of Human Nature

In this section we ask which desiderata a scientifically credible contemporary conception of human nature should seek to fulfil. Kronfeldner and collaborators have distinguished three main epistemic roles for the concept of human nature (Kronfeldner et al. 2014; Kronfeldner Chapter 10 this volume). The first is a definitional or classificatory role: human nature defines the boundary of the human and determines which individuals are members of the human species. The second is a descriptive role: the concept collects the cluster of traits characteristic of the human life form. This can be seen as making human nature an explanandum, something that stands in need of explanation. The second role is therefore complemented by a third role: the concept as an explanans, identifying the underlying mechanisms or factors that explain why humans have this cluster of traits. There is also a fourth, normative role for the concept in answering the question of what a 'typical' or 'proper' human ought to be. While this is one of the most important traditional roles of the concept of human nature, it has few supporters in philosophy of biology (a seminal critique in this field is Hull 1986). Philosophers who still try to use human nature for this normative purpose do not derive their account of human nature from biology (e.g. the neo-Aristotelian accounts reviewed in Glackin 2016). If species had fixed, typical, and teleological natures in the way that folk biology supposes, then human nature could fulfil the first three roles Kronfeldner identifies, and perhaps the fourth. But since there are no such natures, a scientifically credible concept of human nature must be somewhat revisionary. It will give people something of what they originally wanted from a concept of human nature, but not everything.

We believe that a good concept of human nature should fulfil the following desiderata:

1. It should be explanatory and not merely descriptive. One of us has argued elsewhere that a purely descriptive idea of human nature is relatively uncontroversial (Griffiths 2011). After all, there is a range of sciences that deal with humans, and many of these sciences are successful, which implies that one can abstract away from the particularities of individual human lives to discover commonalities. We suggest that in addition a concept of human nature needs to address what causes these commonalities: it needs to fulfil an explanatory role.

2. This leads to our second desideratum: a useful concept should make human nature an object of inquiry in the human sciences: the sciences that deal with human beings as a kind. For example, physiology tries to understand functional processes in the human body; psychology studies the human mind, its underlying processes, and the behavioural characteristics it produces; sociology investigates the human kind in terms of social relations and institutions; cultural anthropology is the comparative study of these matters; and so forth.
3. A third desideratum concerns the relationship between a new conception of human nature and the existing, vernacular conception. The new conception cannot include as defining conditions of human nature all the features that are associated with the vernacular concept. As we have already mentioned, these are essentially independent biological properties that we should not expect to be tightly associated with each other (Griffiths 2011). But there are important properties that some human phenotypes exhibit, and a concept of human nature should recognize this. For example, the fixity of traits can be explained by canalization (Waddington 1942), and the fact that there are canalized traits should be part of our understanding of human nature. Typicality is not a defining feature of human nature, but the fact that there are some typical features of human beings needs to be encompassed by our understanding of human nature. Teleology is today standardly explained via evolutionary adaptation—some features really are there by evolutionary 'design' and others are not—so our understanding of human nature should recognize that our nature is in part the outcome of evolutionary design.[2]
4. Finally, *contra* Edouard Machery (2008), universality is not a desideratum for a concept of human nature (for a critique, see Ramsey 2012, 2013). If the human species is polymorphic, then this is part of the nature of the human species, something we should seek to understand when we study human beings as a kind. Many organisms also exhibit some form of phenotypic plasticity, the evolved ability to respond with different phenotypes to different environments (Gilbert and Epel 2009; Sultan 2015). This too is an important part of the nature of the species in question. In suggesting that the features of human nature must be universal, Machery is responding to a real feature of the vernacular concept of human nature, but one that clashes with what we have learned about biology since Darwin. So our fourth desideratum is that human nature should admit of polymorphism and plasticity.

In summary, then, we propose that a concept of human nature should make human nature something that explains many features of human beings; that it should make

---

[2] One problematic aspects of the teleological way of thinking is its resistance to counter-evidence. The nonexistence of a so-called essential trait among a large number of members of a population can always be explained as the failure of those individuals to realize their proper nature (we thank Tim Lewens for this comment).

human nature an object of enquiry for the human sciences; that it should make room for developmentally fixed and species-typical traits, and for the fact that some traits are the result of evolutionary design; and finally that it should accommodate the fact that humans are diverse and plastic.

Amongst the many accounts of human nature offered by philosophers, two meet these desiderata. The first is Grant Ramsey's life history trait cluster (LTC) account of human nature (Ramsey 2013) and the second is the developmental systems (DS) account of human nature (Stotz 2010; Griffiths 2011). In the next section we explain the similarities and complementarities of these two accounts, and in section 3.5 we turn to the differences between them.

## 3.4 LTC and DST: Human Nature as Human Development

So what is the LTC account? Ramsey acknowledges that human beings are diverse, with each individual life history including a different mix of traits. His account focuses on the patterns of co-occurrence between traits in this population of diverse life histories:

Human nature is defined as the pattern of trait clusters within the totality of extant human possible life histories. Thus, if one were to take all of the possible life histories that form the basis for individual nature, and then combine them, one would possess the set of life histories that forms the basis for human nature, since the trait distribution patterns in this set of life histories constitute human nature. (Ramsey 2013: 987)

Two ideas are combined in this proposal, both of which are central to developmental systems theory: first, 'from an evolutionary point of view an animal is the implementation of a life-history strategy'; second, 'bringing order to that diversity is not about identifying universal elements, but about finding order in the patterns of similarity and difference' (Griffiths 2011: 325, 328). In fact, as we now go on to show, the two accounts are remarkably convergent, albeit arriving at their conclusions from very different starting points.

Ramsey identifies two key desiderata for an account of human nature: that it accord both with scientific practice and with intuitive notions of human nature. The first demands its empirical accessibility as a subject to the human sciences, which is in line with our second desideratum. Ramsey also wants his account of human nature to clarify the related concepts of 'innateness and naturalness' (2013: 986). This requirement has something in common with our third desideratum: that a concept of human nature should shed light on the phenomena of typicality, fixity, and teleology. Ramsey's account also embraces developmental plasticity and diversity, and so meets our fourth desideratum.

There are other similarities between the two accounts. Ramsey eschews any classificatory role for human nature: an organism is human because it is a member of a particular lineage, not because it displays the LTC property clusters. He also eschews a normative role for human nature: his account may illuminate the idea that some traits are 'natural', but it is not intended as an account of how human beings should be. We agree with both of these points.

Ramsey sometimes seems to regard his account as merely descriptive and not explanatory: 'characterizations of features of human nature are merely descriptions of patterns within the collective set of human life histories' (2013: 988). This apparently clashes with our first desideratum, which calls for an explanatory account of human nature. For two reasons, however, we do not see this as a major difference between our accounts. First, Ramsey's life history trait clusters are exactly what our developmental systems account of human nature explains, which makes his account complementary to ours. Second, we believe that Ramsey actually presents an account of how individual traits that make up the trait cluster can be explained by human nature.

Ramsey's account is more than merely descriptive, we think, because it does not simply list features as a description of human nature. The account focuses on the identification of 'antecedent' (A) and 'consequent' (C) traits of life histories that have been found to be associated with each other. Further experiments should then be carried out, Ramsey suggests, to determine if As and Cs are causally related rather than merely correlated. This would amount to an experimental programme to establish constraints on the possible trajectories within life-history space, and hence the beginning of an explanation of human nature, as well as a description.

Ramsey sketches a quasi-formal account of human nature, involving a 'human-nature space'. This is not the state space of human life histories, as in the last paragraph, but a theoretical space in which to locate and compare particular trait clusters. It has two dimensions: the 'pervasiveness, $p$, of the antecedent', defined as 'the proportion of life histories that exhibit that trait', and the 'robustness, $r$, of the antecedent-consequent association' (we are unclear if $r$ is simply a correlation, something like the regression of the consequent on the antecedent, or more explicitly a causal measure). One can increase $p$ by choosing a more broadly defined antecedent, but this will typically reduce the robustness of its association with a consequent. Equally, adding more antecedent traits—make it more complex—can increase $r$, but at a cost to $p$. Hence there is a trade-off between $p$ and $r$, or between simplicity and strength (Ramsey 2013: 989–90).

Ramsey argues that one can make sense of both innateness and naturalness in terms of positions within the $p$–$r$ space. It may be natural, part of human nature, for humans that have property A to also have property C—for example, being female (A) and menstruating (C). Since human nature is also associated with traits being innate, innateness could be interpreted in various ways in terms of the $p$–$r$ space. Either the higher the $r$-value, the more innate a trait is; or innateness can be defined as association with both a high $p$-value and a high $r$-value; or, since neither of these two proposed definitions of innateness implies 'not learned', one could restrict the term 'innate' to A–C links that involve no learning. This, Ramsey (2013: 991) admits, might exclude most, if not all, associations, 'since learning is woven into the causal fabric of so much of development'.

Ramsey (2013: 987) notes that an LTC account of human nature may seem 'spectacularly—and perhaps disastrously—permissive' and 'extremely inclusive'. However, Ramsey argues that although LTC is in principle very permissive, in that it includes all trait associations, it does not imply that all these associations are equally interesting. He proposes the $p–r$ space as a way to distinguish the more interesting features of human nature, those most worthy of study in the science of human nature. Insofar as these interesting trait associations are the ones that are more 'natural' or 'innate', this seems to us to be another residual influence of the folk-biological conception of human nature. In any case, Ramsey does not need to defend himself against this criticism. As we explain in the next section, the idea that accounts of human nature should not be 'permissive' or 'inclusive' is simply mistaken.

Finally, Ramsey claims that while his account is not normative, it nevertheless has 'moral implications'. Since this account gives us robust insight into the human condition, good and bad, it could guide action via desired or unwanted antecedent–consequent associations (Ramsey 2013: 992). That biology can have moral implications in this straightforward way has often been noticed: 'Starving children stunts their growth and ruins their health and this is one reason not to starve them' (Sterelny and Griffiths 1999: 5).

## 3.5 The Developmental Systems Account of Human Nature

The developmental systems account describes human nature in a way very similar to Ramsey. Organisms are fundamentally processes (Griffiths and Stotz 2018)—life cycles—and heredity is the reconstruction of the life cycle using resources that are passed on by previous generations. Some of these resources are genetic, some epigenetic, and some exogenetic—the last term referring to a 'developmental niche' that contains reliable developmental resources from outside the organisms.[3] Some exogenetic resources serve to canalize development, and some to modify it and hence enable developmental plasticity.

Developmental systems theorists have long recognized Ramsey's point that a single lineage has many possible developmental trajectories: 'life cycles may have a disjunctive form, with different individuals having different characteristics. A developmental system can proliferate by producing a range of outcomes on different occasions' (Griffiths and Gray 1994: 296). Descriptive human nature is the 'order in the patterns of similarity and difference' in these human life cycles (Griffiths 2011: 328).

Developmental systems theory explains human nature as the product of the human developmental system, a matrix of genetic, epigenetic, and exogenetic resources within which the developmental process or life cycle unfolds. This system is constructed by

---

[3] For an account of the unique features of human nature that accords a very substantial role to the developmental niche, see Sterelny (2003, 2012).

earlier human life cycles and by feed-forward effects from the development of the individual itself. Progress in understanding human nature, on this view, is simply progress in the sciences of human development: developmental biology, developmental psychobiology, and developmental psychology (for a brief history of DST and the scientific research traditions from which it emerged, see Griffiths and Tabery 2013).

Ramsey's fear that the LTC account might be criticized as 'spectacularly—and perhaps disastrously—permissive' echoes the assessment by Kronfeldner and collaborators of the developmental systems account of human nature:

The result is a concept of an *all-inclusive* human nature that comprises all the resources needed to stabilize the development of the patterns of similarity and difference observable in humankind. Human nature, the thing that explains and defines the human species, is then a genealogically anchored explanatory essence of *gigantic proportions*, namely the *whole* developmental system of humankind, including the developmental niche [...this is] a very distant relative of the traditional concept of human nature, since it construes everything involved in and resulting from human development as part of human nature. It is doubtful whether such an all-encompassing concept of human nature is of any concrete use for the sciences, that is, for describing and explaining commonalities or explaining differences within humankind or between the human and other species.

(Kronfeldner et al. 2014: 649; emphasis added)

This is a non sequitur: the observations made about the DST account do not support the conclusion. In fact, the DST account has a better prospect of 'describing and explaining commonalities or explaining differences' than does the 'traditional concept of human nature', which excludes much of human diversity. If the aim was to pick out some individuals as not human or as less human, then we might need a simple definition of the human, like a CO1 gene barcode, but hopefully no one is trying to do that! If, instead, the aim of studying human nature is to understand what human beings are like and why they are like that, then we see no reason why either the description of human nature or its underlying explanation should be simple. It seems obvious that both will be complex.

Kronfeldner et al. seem to be echoing a common criticism levelled at developmental systems theory: that paying attention to the role of the environment in development and to the plasticity of development will make the study of development scientifically intractable and its results incomprehensibly complex. The same accusation has been levelled against the scientists whose work inspired DST, and the reply is the one those scientists gave—'development *is* complicated' (Bateson 1991: 19). On our view, and on Ramsey's view as we interpret it, complex interactions between genetic, epigenetic, and exogenetic factors explain the constraints on developmental trajectories in the state space of possible human life histories that constitute human nature.[4] It would be

---

[4] A difference between Ramsey's account and our own may concern the status of the genome, which DST put on a much more equal footing to the environment than it enjoys on the LTC account: 'If genes were allowed to vary, individual nature would be vacuous since sufficient changes to genes could, say, change an American into an aardvark. By contrast, varying the way that an individual encounters its environmental heterogeneity reveals something about its nature' (Ramsey, Ch. 2 this volume).

convenient if these could be reduced to a few simple parameters, like the average velocity of molecules in a gas, but it is clear that they cannot.[5]

Developmental systems theorists have repeatedly emphasized that an inclusive definition of the developmental system does not mean that the whole system must be studied at once, any more than the inclusive definition of the proteome precludes studying individual protein–protein interactions (Griffiths and Gray 2005; Oyama 2000). The concept of the developmental niche, which seems to be of particular concern to Kronfeldner et al., is a construct from empirical research on behavioural development (West and King 1987). It was introduced into DST to give greater structure to the extra-organismic component of the developmental system (Stotz and Allen 2012; Stotz and Griffiths 2016). The developmental niche concept has been used to great effect in such different fields as the development of social behaviour and communication in birds (West and King 1987, 2008), and species-typical development in general in rats (Alberts 2008). Other research groups have applied DST's view of development and the concept of the developmental niche to investigate aspects of human development (Alberts and Ronca 2012; Gros-Louis et al. 2014; Gros-Louis et al. 2016; Narvaez et al. 2013). None of this research has become mired in an unmanageable sea of complexity because it recognizes that the human life cycle has evolved to make use of a highly specific developmental niche, or that interaction with this niche may induce developmental plasticity.

Developmental systems theory does not make it possible to sum up human nature in a slogan, but it does point clearly to the body of knowledge that constitutes our current best understanding of human nature: human developmental biology, developmental psychobiology, and developmental psychology. When those sciences are complete, we will have a complete understanding of human nature. We fail to see the force of the objection 'but that will be very complicated'.

## 3.6 A Distinctive Feature of the DS Account: Human Developmental Niche Construction

While Ramsey focuses on descriptive property clusters that make up human nature, the developmental systems account focuses on the underlying processes that account for these clusters. Developmental systems theory subscribes to a process account of the organism, and this is reflected in its view of human nature. DST is a process theory because developmental systems are essentially extended in time (Griffiths and Gray 1994, 1997; Griffiths and Stotz 2018):

[DST] seeks to explain developmental outcomes as the result of a dynamic process in which some of the interacting factors are products of earlier stages of the process, rather than as the

---

[5] This is not to reject research programmes in systems biology that aim at substantial reductions in the complexity of development through identifying systems-level variables.

result of the arrangement of pre-existing factors into a static mechanism. Even when factors exist independently of the developmental process, they are drawn into it and made part of a developmental 'system' by the unfolding process. (Griffiths and Stotz 2018)

The focus on property clusters makes Ramsey's LTC account look less processual. However, these properties are not merely properties of organisms, but properties of an organism at a time, and the property clusters that constitute human nature are correlations between what happens at one point in a life cycle and what happens at a later point. So Ramsey's account actually fits a process view of the organism quite well. Moreover, Ramsey conceives of the series of events that make up an individual human being as a life history, the implementation of an evolved strategy for resource allocation across the lifespan. In our recent work (Griffiths and Stotz 2018) we have argued that it is a life-history strategy that constitutes the principle of identity which unites a series of events as a single life cycle, rather than a part of a larger cycle, or a process involving more than one individual.

The main difference between the two accounts is that the DS account has a stronger focus on the role of the environment in constituting human nature. There is an old saying within anthropology that culture is not only part of human nature, but that our nature is culture. Some recent work on human evolution has emphasized the role of selective niche construction: the evolution of the unique characteristics of human psychology and social structure has been substantially driven by the selection pressures created by earlier psychologies and social structures (Laland et al. 2000; Sterelny 2012). Niche construction theory deals with the selective niche, defined by the parameters that determine the relative fitness of competing types in a population. In selective niche construction, earlier generations partly construct the selection pressures that act on future generations. But another aspect of human niche construction is that our development is dependent on a rich developmental niche of interaction with parents and other conspecifics, and with physical and cognitive artefacts from tools to languages. The developmental niche is defined by the parameters needed to ensure the reconstruction of the evolved life cycle. The concept of the developmental niche is designed to integrate and formalize the non-genetic yet heritable factors influencing an organism's development (Stotz 2010, 2014, 2017; Griffiths and Stotz 2013). It was first proposed under the name 'ontogenetic niche' by developmental psychobiologists West and King (1987).

In our current formulation of the concept (Griffiths and Stotz 2013, 2018), the developmental system consists of genetic resources, epigenetic resources, and an exogenetic developmental niche, which contains reliably inherited physical, social, ecological, and epistemic resources needed to reconstruct or—in the case of phenotypic plasticity—modify that developmental system. These resources can be actively constructed by the parents (producing the 'parental effects' of quantitative genetics) or by the larger group, co-constructed by parent and offspring, or sourced passively from a stable environment. Wherever they come from, if there exists an evolutionary (historical) explanation for the interaction of the evolved developmental system with the resource, then that resource is part of the system. What evolves by natural selection is a relationship between system and each resource.

How does the developmental niche influence human development? Human babies are needy. They are born early in comparison to other primates, meaning that for several months postnatally, relative to other primates, human babies share characteristics of foetuses rather than of infants in those other primates (Trevathan 2011). Comparing brain size at birth among primates, humans should be born at 18 months of age. A large part of brain development takes place outside the uterus, allowing for much greater postnatal epi- and exogenetic influence than for their ape cousins, which makes the early niche fundamental for human development. Over the course of human evolution, as brains became bigger and human infants more immature at birth, human childrearing practices evolved in tandem to ensure the survival of the helpless infant. As bipedalism, hemochorial placenta, large brains, and the need for a greater amount of learning after birth emerged, human evolution intensified parental care: 'Only with intensified parental care in response to greater helplessness of the infant could selection favor the evolution of a large brain in a bipedal animal' (Trevathan 2011: 33). So the emergence of a more complex and resource-demanding developmental niche has been a key feature of human evolution.

For this reason, it seems to us entirely natural to say that that human nature resides partly in the human developmental environment. We are a species that is particularly strongly influenced by niche construction, both selective niche construction over evolutionary timescales and developmental niche construction over ontogenetic timescales. A concept of nature according to which what is natural must come from the inside is particularly unsuitable for such a species. Imagine trying to determine the real nature of an ant, another powerful niche constructor, by removing the influence of the nest on the developing egg and embryo. The result would be either dead or biologically meaningless; and so it is for humans.

The developmental niche has two fundamental functions. One function is to ensure the stable, reliable development of species-typical traits. So what explains typicality is the developmental systems dynamics within what we may call 'normal' parameters, some of which are provided by pre-existing physical and developmental constraints. The rest are ensured by reliably and stably inherited resources, which include not just the genome but also essential environmental resources that (among other functions) assist in the species-typical expression of the genetic factors. These stable resources also partially explain fixity. In addition, there are developmental mechanisms that buffer against internal (genetic, epigenetic, metabolic) and external perturbations. These are invoked when we talk about canalization.

But one of our desiderata was that human nature needs to embrace and explain human diversity: 'The search for a shared human nature cannot be the search for human universals; it must instead be a way to interpret and make sense of human diversity' (Griffiths 2011: 326). Here the second function of the developmental niche comes in. Beyond ensuring reliable development, the developmental niche also provides input to developmental plasticity. Plasticity is often defined in terms of a genotype's ability to produce different phenotypes in response to the environment. It would be more accurate, however, to say that the shape of the norm of reaction is a property of the whole developmental system. So what explains human diversity are differing developmental systems dynamics supported by modifications in the developmental niche. In other words, human diversity results primarily from the interaction between the evolved developmental system and a wide range of environments, including novel environments: 'Bringing order to that diversity is not about identifying universal elements, but about finding order in the patterns of similarity and difference' (Griffiths 2011: 328). Developmental niche construction therefore provides dependability, but also adaptive flexibility, in the provision of necessary developmental resources.

## 3.7 Conclusion

In this chapter we have reiterated our view that human nature is simply human development. To the extent that we understand human developmental biology, developmental psychobiology, and developmental psychology, we understand human nature. Like Ramsey's LTC account of human nature, this amounts to saying that human nature is a set of constraints on possible human developmental trajectories. Like Ramsey's account, it is not without content because, although it does not identify a set of outcomes that are 'unnatural', it does say that 'you can't get there from here'. This gives our account, like Ramsey's, a very special and positive feature: it is able to embrace human diversity as part of human nature. As we have argued in section 3.2, the objection that our account leads to a very complex picture of human nature is a non sequitur: human nature *is* complicated.

Our account differs from Ramsey's in a greater stress on the role of the human developmental environment—the developmental niche—in constituting human nature. We have argued that this reflects the direction of the human sciences in recent years. If it clashes with a folk-biological intuition that nature must come from 'inside' rather than 'outside', so much the worse for that intuition—we understand human nature better now.

Kronfeldner and colleagues (Kronfeldner et al. 2014; Kronfeldner Chapter 10 this volume) have proposed a 'pluralistic solution' for the missing consensus in the philosophical literature regarding a concept of human nature. They suggest that 'different scientific fields are in need of different concepts of human nature, each fulfilling an independent epistemic role'. We are sympathetic to this general approach to the analysis of scientific concepts. We ourselves have made a similar suggestion about the concept of the gene: different gene concepts should be understood as 'tools of research, as ways of classifying the experience shaped by experimentalists to meet their specific needs' (Stotz and Griffiths 2008: 41; see also Griffiths and Stotz 2013). We do not, however, think that the same argument applies to the concept of human nature. Human nature is less a technical concept applied in the laboratory than a pragmatic, and even normative, tool applied in wider social contexts and with wide-ranging consequences. This does not mean that different scientific endeavours cannot study different aspect of human nature, but they cannot do this without paying attention to other fields. There have been several attempts to impose a simplistic understanding of human nature, often derived from evolutionary biology, and to marginalize other sciences, such as those that focus on the human developmental environment. As Sandra Mitchell has argued, an 'anything goes' pluralism in science may do more harm than good, while a real 'integrative pluralism' is a useful defence against reductionist imperialism (Mitchell 2003, 2009).

# References

Alberts, J. R. (2008). 'The Nature of Nurturant Niches in Ontogeny.' *Philosophical Psychology* 21: 295–303.

Alberts, J. R., and Ronca, A. E. (2012). 'The Experience of Being Born: A Natural Context for Learning to Suckle.' *International Journal of Pediatrics* e129328: doi:10.1155/2012/129328.

Atran, S. (1990). *Cognitive Foundations of Natural History: Towards an Anthropology of Science*. Cambridge: Cambridge University Press.

Bateson, P. (1991). 'Are There Principles of Behavioural Development?' In P. Bateson (ed.), *The Development and Integration of Behaviour: Essays in Honour of Robert Hinde*, 19–39. Cambridge: Cambridge University Press.

Berlin, B. (1992). *Ethnobiological Classification: Principles of Classification of Plants and Animals in Traditional Societies*. Princeton, NJ: Princeton University Press.

Cheung, B. Y., Dar-Nimrod, I., and Gonsalkorale, K. (2014). 'Am I My Genes? Perceived Genetic Etiology, Intrapersonal Processes, and Health.' *Social and Personality Psychology Compass* 8: 626–37.

Dar-Nimrod, I., and Heine, S. J. (2011a). 'Genetic Essentialism: On the Deceptive Determinism of DNA.' *Psychological Bulletin* 137: 800–818.

Dar-Nimrod, I., and Heine, S. J. (2011b). 'Some Thoughts on Essence Placeholders, Interactionism, and Heritability: Reply to Haslam (2011) and Turkheimer (2011).' *Psychological Bulletin* 137: 829–33.

Dar-Nimrod, I., and Lisandrelli, G. (2012). 'It's in My Genes: Causal Attributions and Perceptions of Choice.' *Advances in Psychology Research* 95: 187–98.

Dar-Nimrod, I., Zuckerman, M., and Duberstein, P. R. (2012). 'The Effects of Learning about One's Own Genetic Susceptibility to Alcoholism: A Randomized Experiment.' *Genetics in Medicine* 15: 132–8.

Dar-Nimrod, I., Zuckerman, M., and Duberstein, P. (2014). 'Smoking at the Workplace: Effects of Genetic and Environmental Causal Accounts on Attitudes toward Smoking Employees and Restrictive Policies.' *New Genetics and Society* 33: 400–412.

Downes, S. M., and Machery, E. (2013). *Arguing about Human Nature: Contemporary Debates*. New York: Routledge.

Fuentes, A., Marks, J., Ingold, T., and Sussman, R. (2010). 'On Nature and the Human.' *American Anthropologist* 112: 512–21.

Gelman, S. A. (2003). *The Essential Child: Origins of Essentialism in Everyday Thought*. New York: Oxford University Press.

Gilbert, S., and Epel, D. (2009). *Ecological Developmental Biology: Integrating Epigenetics, Medicine, and Evolution*. Sunderland, Mass.: Sinauer Associates.

Glackin, S. N. (2016). 'Three Aristotelian Accounts of Disease and Disability.' *Journal of Applied Philosophy* 33: 311–26.

Gottlieb, G. (1970). 'Conceptions of Prenatal Behavior.' In L. R Aronson, E. Tobach, D. S. Lehrman, and J. S. Rosenblatt (eds), *Development and Evolution of Behavior: Essays in Memory of T. C. Schneirla*, 111–37. San Francisco, Calif.: W. H. Freeman.

Griffiths, P. E. (2002). 'What Is Innateness?' *The Monist* 85: 70–85.

Griffiths, P. E. (2011). 'Our Plastic Nature.' In S. Gissis and E. Jablonka (eds), *Transformations of Lamarckism: From Subtle Fluids to Molecular Biology*, 319–30. Cambridge, Mass.: MIT Press.

Griffiths, P. E., and Gray, R. D. (1994). 'Developmental Systems and Evolutionary Explanation.' *Journal of Philosophy* 91: 277–304.

Griffiths, P. E., and Gray, R. D. (1997). 'Replicator II: Judgment Day.' *Biology and Philosophy* 12: 471–92.

Griffiths, P. E., and Gray, R. D. (2005). 'Three Ways to Misunderstand Developmental Systems Theory.' *Biology and Philosophy* 20: 417–25.

Griffiths, P. E., and Machery, E. (2008). 'Innateness, Canalisation, and "Biologicizing the Mind".' *Philosophical Psychology* 21: 397–414.

Griffiths, P. E., Machery, E., and Linquist, S. (2009). 'The Vernacular Concept of Innateness.' *Mind and Language* 24: 605–30.

Griffiths, P. E., and Stotz, K. (2013). *Genetics and Philosophy: An Introduction*. New York: Cambridge University Press.

Griffiths, P. E., and Stotz, K. (2018). 'How DST Became a Process Theory.' In D. J. Nicholson and J. Dupré (eds), *Everything Flows: Towards a Processual Philosophy of Biology*. Oxford: Oxford University Press.

Griffiths, P. E., and Tabery, J. G. (2013). 'Developmental Systems Theory: What Does It Explain, and How Does It Explain It?' In R. M. Lerner and J. B. Benson (eds), *Embodiment and Epigenesis: Theoretical and Methodological Issues in Understanding the Role of Biology within the Relational Developmental System, Part A: Philosophical, Theoretical, and Biological Dimensions*, 65–94. New York: Academic Press.

Gros-Louis, J., West, M. J., and King, A. P. (2014). 'Maternal Responsiveness and the Development of Directed Vocalizing in Social Interactions.' *Infancy* 19: 385–408.

Gros-Louis, J., West, M. J., and King, A. P. (2016). 'The Influence of Interactive Context on Prelinguistic Vocalizations and Maternal Responses.' *Language Learning and Development* 12: 280–94.

Hinde, R. A. (1968). 'Dichotomies in the Study of Development.' In J. M. Thoday and A. S. Parkes (eds), *Genetic and Environmental Influences on Behaviour*, 3–14. New York: Plenum.

Hull, D. L. (1986). 'On Human Nature.' *Proceedings of the Philosophy of Science Association* 1: 3–13.

Jayaratne, T. E., Gelman, S. A., Feldbaum, M., Sheldon, J. P., Petty, E. M., and Kardia, S. L. R. (2009). 'The Perennial Debate: Nature, Nurture, or Choice? Black and White Americans' Explanations for Individual Differences.' *Review of General Psychology* 13: 24–33.

Kronfeldner, M., Roughley, N., and Toepfer, G. (2014). 'Recent Work on Human Nature: beyond Traditional Essences.' *Philosophy Compass* 9: 642–52.

Laland, K. N., Odling-Smee, F. J., and Feldman M. W. (2000). 'Niche Construction, Biological Evolution, and Cultural Change.' *Behavioral and Brain Sciences* 23: 131–57.

Lehrman, D. S. (1953). 'Critique of Konrad Lorenz's Theory of Instinctive Behavior.' *Quarterly Review of Biology* 28: 337–63.
Lewens, T. (2012). 'Human Nature: The Very Idea.' *Philosophy and Technology* 25: 459–74.
Linquist, S., Machery, E., Griffiths, P. E., and Stotz K. (2011). 'Exploring the Folkbiological Conception of Human Nature.' *Philosophical Transactions of the Royal Society B* 366: 444–53.
Machery, E. (2008). 'A Plea for Human Nature.' *Philosophical Psychology* 21: 321–9.
Mameli, M., and Bateson, P. P. G. (2006). 'Innateness and the Sciences.' *Biology and Philosophy* 21: 155–88.
Mameli, M., and Bateson, P. P. G. (2011). 'An Evaluation of the Concept of Innateness.' *Philosophical Transactions of the Royal Society of London B* 366: 436–43.
Medin, D. L., and Atran, S. (1999). 'Introduction.' In Medin and Atran (eds), *Folkbiology*, 1–15. Cambridge, Mass.: MIT Press.
Medin, D., and Atran, S. (2004). 'The Native Mind: Biological Categorization and Reasoning in Development and across Cultures.' *Psychological Review* 111: 960–83.
Medin, D., and Ortony, A. (1989). 'Psychological Essentialism.' In S. Vosniadou and A. Ortony (eds), *Similarity and Anological Reasoning*, 175–95. Cambridge: Cambridge University Press.
Mitchell, S. D. (2003). *Biological Complexity and Integrative Pluralism*. Cambridge: Cambridge University Press.
Mitchell, S. D. (2009). *Unsimple Truths: Science, Complexity, and Policy*. Chicago: University of Chicago Press.
Narvaez, D., Gleason, T., Wang, L. J., and Brooks, J. (2013). 'The Evolved Development Niche: Longitudinal Effects of Caregiving Practices on Early Childhood Psychosocial Development.' *Early Childhood Research Quarterly* 28: 759–73.
Oyama, S. (2000). *Evolution's Eye: A Systems View of the Biology–Culture Divide*. Durham, NC: Duke University Press.
Parrott, R., and Smith, R. A. (2014). 'Defining Genes Using "Blueprint" versus "Instruction" Metaphors: Effects for Genetic Determinism, Response Efficacy, and Perceived Control.' *Health Communication* 29: 137–46.
Ramsey, G. (2012). 'How Human Nature Can Inform Human Enhancement: A Commentary on Tim Lewens's "Human Nature: The Very Idea".' *Philosophy and Technology* 25: 479–83. doi: 10.1007/s13347-012-0087-2.
Ramsey, G. (2013). 'Human Nature in a Post-Essentialist World.' *Philosophy of Science* 80: 983–93.
Sterelny, K. (2003). *Thought in a Hostile World: The Evolution of Human Cognition*. Oxford: Blackwell.
Sterelny, K. (2012). *The Evolved Apprentice*, Cambridge, Mass.: MIT Press.
Sterelny, K., and Griffiths, P. E. (1999). *Sex and Death: An Introduction to the Philosophy of Biology*. Chicago: University of Chicago Press.
Stotz, K. (2010). 'Human Nature and Cognitive-Developmental Niche Construction.' *Phenomenology and the Cognitive Sciences* 9: 483–501.
Stotz, K. (2014). 'Extended Evolutionary Psychology: The Importance of Transgenerational Developmental Plasticity.' *Frontiers in Psychology* 5: doi: 10.3389/fpsyg.2014.00908.
Stotz, K. (2017). 'Why Developmental Niche Construction Is Not Selective Niche Construction—and Why It Matters.' *Interface Focus*, 7: 2–10.

Stotz, K., and Allen, C. (2012). 'From Cell-Surface Receptors to Higher Learning: A Whole World of Experience'. In K. Plaisance and T. Reydon (eds), *Philosophy of Behavioural Biology*, 85–123. Boston, Mass.: Springer.

Stotz, K., and Griffiths, P. E. (2008). 'Biohumanities: Rethinking the Relationship between Biosciences, Philosophy and History of Science, and Society'. *Quarterly Review of Biology* 83: 37–45.

Stotz, K., and Griffiths, P. E. (2016). 'A Niche for the Genome'. *Biology and Philosophy* 31: 143–57.

Sultan, S. E. (2015). *Organism and Environment: Ecological Development, Niche Construction, and Adaptation*. Oxford: Oxford University Press.

Trevathan, W. (2011). *Human Birth: An Evolutionary Perspective*. New York: Aldine de Gruyter.

Waddington, C. H. (1942). 'Canalisation of Development and the Inheritance of Acquired Characters'. *Nature* 150: 563–5.

West, M. J., and King, A. P. (1987). 'Settling Nature and Nurture into an Ontogenetic Niche'. *Developmental Psychobiology* 20: 549–62.

West, M. J., and King, A. P. (2008). 'Deconstructing Innate Illusions: Reflections on Nature-Nurture-Niche from an Unlikely Source'. *Philosophical Psychology* 21: 383–95.

# 4
# Human Nature, Natural Pedagogy, and Evolutionary Causal Essentialism

*Cecilia Heyes*

## 4.1 Introduction

The concept of human nature is used not only by academics but a great deal in private and public life. Like Agatha Christie's Miss Marple, many people regard their day-to-day interest in the activities and motivations of others as an interest in human nature, and it is common for public figures—journalists, judges, politicians, and public intellectuals of all stripes—to invoke human nature when discussing aspects of our behaviour or mentality that they regard as important, deep-seated, or immutable. Indeed, the concept of human nature has so much currency outside academic circles that it is hard to believe its abandonment by scientists and philosophers of science—no matter how compelling the reasons—would prompt a similar move in the outside world. Rather, if the concept of human nature is eliminated from scientific discourse, it will almost certainly continue to be used 'out in the world', but that usage will not be informed or constrained by science; an important channel of communication between science and the public will close down. For this reason, and because the concept of human nature plays an important role in defining explanatory projects within science (Sterelny, Chapter 6 this volume), I believe it should be patched up rather than eliminated.

This chapter discusses the 'theory of natural pedagogy': an account of how genetic evolution has made human infants receptive to teaching. This theory does not appeal explicitly to the concept of human nature. However, the inclusion of 'natural' in its name, and the purpose of the theory—to describe universal and distinctively human cognitive mechanisms—make the theory of natural pedagogy representative of the widely held view that human nature is rooted in genetic evolution. Challenging this view, I argue that empirical research relevant to the theory of natural pedagogy—and, viewed from a certain angle, the theory itself—highlight the importance of cultural evolution in shaping human nature. At the end, I outline an 'evolutionary causal essentialist' way of patching up the concept of human nature that, among other assets, recognizes the importance of cultural evolution in shaping distinctively human cognitive mechanisms (Heyes 2012).

## 4.2 Natural Pedagogy

The theory of natural pedagogy has been an important focus of research on the development and evolution of human cognition for ten years (Csibra and Gergely 2006; Gergely and Csibra 2005). It offers an account of how teaching evolved, and of the psychological processes that make infants and children receptive to teaching. In essence, the theory proposes that human infants genetically inherit a 'well-organised package of biases, tendencies and skills' (Csibra and Gergely 2006: 8) making them receptive to deliberate attempts by adults to convey information. This package constitutes a biological adaptation for teaching: it was favoured by natural selection operating on genetic variants because it enhanced the fidelity of cultural inheritance (Csibra and Gergely 2011).

When summarizing the components of the natural pedagogy package, Csibra and Gergely use three headings: ostension, reference, and relevance.

> *Ostension*: Infants' sensitivity to *eye contact, contingencies,* and *infant-directed speech* make them more likely to learn by observing information that the adults intend the infants to learn.
>
> *Reference*: In the presence of ostensive or communicative cues (i.e. when an adult is making eye contact, responding contingently, and/or using infant-directed speech), infants tend to shift their attention in the direction indicated by the adult's gaze. This *gaze cuing* increases the probability that the infant will learn about the object or event that the adult intends the infant to learn about.
>
> *Relevance*: In the presence of ostensive cues, in a 'communicative context', infants are more likely to copy features of an adult's behaviour that are opaque to them than features that the infant can already understand. This *rational imitation* bias increases the probability that, as the teacher intends, the infant will acquire through observation of the teacher's behaviour information that is new to them, and that can be used in a range of contexts.

It is sometimes unclear whether Csibra and Gergely are proposing that infants are receptive to teaching by virtue of low-level sensorimotor processes or high-level inferential processes. Consequently, their theory has been criticized both for being too lean—placing too much emphasis on automatic psychological processes (Nakao and Andrews 2014)—and for being too rich—attributing to infants inferential feats that are likely to be beyond their cognitive power (Beisert et al. 2012). In this chapter and elsewhere (Heyes 2016), I take it that the theory of natural pedagogy assumes that low-level mechanisms make young infants receptive to teaching, and that when describing infants' competence in high-level terms—using words such as 'believe', 'conceive', 'infer', and 'rational'—the authors of the theory are adopting the intentional stance (Dennett 2009: 339). In other words, I take it that Csibra and Gergely use intentional terms as a way of presenting their hypotheses about the adaptive functions of infants' competence, not about the psychological mechanisms underlying that competence. This view is consistent with Csibra and Gergely having stated explicitly that many components of the natural pedagogy package can be mediated by low-level processes, their denial that natural pedagogy depends on language or theory of mind (Csibra and Gergely 2006), and references to Dennett's intentional stance in their previous work (Gergely et al. 1995).[1]

In a recent article (Heyes 2016), I reviewed evidence from studies of infants, children, adults, and non-human animals relating to each of the five principal components of the natural pedagogy package: eye contact, contingencies, infant-directed speech, gaze cuing, and rational imitation. In each case, the review asked whether there is compelling evidence that the component is (1) real, (2) a genetic adaptation, and (3) adapted specifically to promote teaching. A component was judged to be real when it had been shown to be a reliable feature of infant minds and behaviour. Following the lead of Gergely and Csibra, and other developmental psychologists who are interested in evolution, a component was taken to be a genetic adaptation when there was evidence that it is present at birth (or 'inborn') and/or that its development depends on specialized, rather than domain-general, mechanisms of learning. Finally, whether a component is an adaptation for teaching (3) was addressed by task analysis—asking whether there are other functions it could subserve—and by looking at the phylogenetic distribution of the trait; specifically, whether the component is present in species where teaching has not been observed. The conclusions of the review are summarized in Table 4.1. It found:

---

[1] Csibra and Gergely's views may well have evolved in the last decade, such that they are now more inclined to believe that high-level mechanisms implement natural pedagogy. If so, treating these innovations as part of the theory would make the theory more vulnerable, because it would add to claims about the adaptive function of pedagogy a set of hypotheses about the psychological mechanisms that implement those functions (Dennett 2009). I believe that the original version of the theory remains coherent, interesting, and ripe for further development as a functional account. Therefore this chapter does not target any additional—and, in my view, weaker—claims about high-level psychological mechanisms.

Table 4.1 Summary of the evidence from infants, children, adults and non-human animals relating to four questions about the components of natural pedagogy (eye contact, contingencies, infant-directed speech, gaze cuing, rational imitation). For each component, the empirical review presented in Heyes (2016) asked: (1) Is it a reliable phenomenon? (2) Is it a genetic adaptation? (3) Is it a genetic adaptation specifically for teaching, rather than social bonding or social learning? (4) Is it cultural adaptation for teaching, produced by domain-general mechanisms of learning through social interaction?

|  | Reliable phenomenon | Genetic adaptation | Genetic adaptation for teaching | Cultural adaptation for teaching |
|---|---|---|---|---|
| Eye contact | ✓ | ? | ✗ | ? |
| Contingencies | ✓ | ✓ | ✗ | ? |
| Infant-directed speech | ✓ | ✓ | ✗ | ✓ |
| Gaze cuing | ✓ | ✗ | ✗ | ✓ |
| Rational imitation | ? | ✗ | ✗ | ✓ |

*Eye contact*: Current evidence confirms that humans are, from a very young age, highly sensitive to whether another agent is looking at them directly. However, it is possible that, rather than being present at birth, sensitivity to eye contact is a rapidly developing consequence of an inborn preference for face-like stimuli. If eye-contact sensitivity is, in this sense, secondary to a face preference, it may or may not be a genetic adaptation. It is possible that natural selection favoured the genes that promote a face preference in environments where a face preference did not support the early development of eye-contact sensitivity, or where the development of eye-contact sensitivity, via domain-general mechanisms of learning, did not have a positive impact on fitness. Moreover, evidence that our sensitivity to eye contact is shared with a wide range of non-human animals suggests that even if it is a primary genetic adaptation, its function is not specific to teaching. It is possible that, in humans, natural selection has amplified the inborn salience of eye contact, but quantitative change does not appear to be what the theory of natural pedagogy is proposing, and currently there is no evidence that it has occurred.

*Contingencies*: There is stronger evidence that humans have an inborn attraction to response-contingent stimulation—for example, to movements of an adult face or a toy that are correlated rather than uncorrelated with the infant's movements. However, data on imprinting in precocial birds suggest that, like eye-contact sensitivity, this component of natural pedagogy has deep evolutionary roots, and has not been invented by natural selection specifically to make human infants receptive to teaching.

*Infant-directed speech*: Similarly, human infants appear to have an inborn preference for high-pitched, emotional speech, but evidence that this preference is also present in non-human animals suggests that, if it is a genetic adaptation, its function is to promote social bonding and coordination rather than a kind of learning that supports cumulative cultural inheritance.

*Gaze cuing*: Motion cuing—a tendency to track movement with the eyes or head—seems to be an inborn, evolutionarily conserved trait. However, the current evidence suggests that the development of gaze cuing in humans—a specific or exaggerated tendency to follow the movement of eyes—is powered by domain-general mechanisms of learning that detect predictive relationships between eye movements and the location of valuable objects and events. These mechanisms may also detect that eye movements are especially good predictors of value when they have been preceded by eye contact and name calling, and thereby support the modulation of gaze cuing by eye contact and infant-directed speech.

*Rational imitation*: It is not clear whether imitation in infancy and childhood is reliably modulated by communicative context (e.g. eye contact, infant-directed speech) and opacity (the degree to which the observed behaviour is comprehensible in the light of the infant's previous experience). Even if imitation is rational in this evolutionary sense—i.e. in the sense that this modulation is likely to be adaptive in typical ecological circumstances—there is no evidence that modulation by communicative context and opacity develops so early that it is more likely to be a genetic adaptation than a product of domain-general learning. In either case, the modulation of imitation by communicative context and opacity may function to facilitate teaching, to increase the probability that a novice will learn what an expert model intends the novice to learn. However, unless rational imitation can trump or overwrite individual learning when the two are in conflict, it has limited potential to mediate cumulative cultural inheritance.

Thus, it looks as if eye-contact sensitivity, attraction to response-contingent stimulation, a preference for infant-directed speech, and gaze cuing are real phenomena, but it is not clear whether imitation is rational. There is evidence that two of the five components of natural pedagogy are inborn—attraction to response-contingent stimulation and a preference for infant-directed speech—and therefore prima facie evidence that these two are genetic adaptations. However, like eye-contact sensitivity and gaze cuing, attraction to response-contingent stimulation and a preference for infant-directed speech have operating characteristics and a phylogenetic distribution suggesting that their functions relate to social bonding and/or social learning, not specifically to teaching.

This picture clearly does not support the view that natural pedagogy is a set of genetic adaptations for teaching. Nonetheless, I believe the theory of natural pedagogy contains three fundamental insights about the psychology of cumulative cultural inheritance: imitation is not enough; the extra comes not only from smart thinking, but also from blind trust; and tweaking is a powerful source of cognitive change. The sections that follow discuss each of these insights in greater depth than in Heyes (2016), explaining why they are important in relation not only to teaching but to all forms of cultural learning.

## 4.3 Imitation is not Enough

The groundwork for much contemporary research on cultural evolution was laid in the 1970s and 1980s by researchers with backgrounds in anthropology, biology, and mathematics (e.g. Boyd and Richerson 1988; Cavalli-Sforza et al. 1982). Echoing the views of many psychologists who had previously assumed a special relationship between imitation and culture (e.g. Washburn 1908; Piaget 1962; Bruner 1978), they suggested that imitation is the cultural analogue of the mechanisms of genetic inheritance; it allows cultural traits—contributing to skills, practices, institutions, and languages—to be passed down from one generation to the next with sufficient fidelity to allow improvements to accumulate over time. In the 1970s and 1980s, it was widely believed that imitation is a distinctively human capacity, or one that humans share only with other great apes. Since it was (and is) also widely assumed that cumulative cultural evolution is distinctively human, this fitted well with the idea that imitation is the primary, or perhaps the only, form of cultural learning. However, in the late 1980s and early 1990s, evidence began to emerge that not only apes but a broad range of non-human animals are capable of imitation, and other doubts about the sufficiency of imitation began to be raised (Custance et al. 1995; Heyes 1993). Csibra and Gergely are not alone in having responded to these developments by rethinking the relationship between imitation and culture, but their voices are the freshest and most radical. Their theory of natural pedagogy retains a role for imitation, but states clearly that imitation—or what they sometimes call 'blind imitation' (Csibra and Gergely 2006)—is not enough. They make a persuasive case that to understand how cumulative cultural evolution is possible, we need to think harder about the receptive side of teaching, about the ways in which novices derive information, not from experts who are going about their business oblivious to the novices' needs, but from experts who are striving to inform them.

Although I think that Csibra and Gergely are right to insist that imitation is insufficient, I am not entirely convinced by their reasons for this conclusion. They suggest that imitation ceased to be an efficient means of cultural learning with the emergence of 'mediated tool use'—the use of objects, not merely to obtain intrinsically attractive outcomes, such as access to the soft flesh inside an animal hide, but as tools to make other tools, or 'second-order tools'. According to Csibra and Gergely's teleological theory (Gergely and Csibra 2003), a precursor to their theory of natural pedagogy, infants and children are prepared by genetic evolution to learn about actions by identifying each action's purpose or goal, and, crucially, to search for goals among the action's intrinsically attractive effects. Since the actions involved in making second-order tools do not have intrinsically attractive effects, the teleological theory suggests that they present a 'learnability problem'; this problem cannot be overcome by blind imitation—by copying a model's actions on an object without understanding their goals—because a blind imitator 'would not know what conditions are appropriate to use the tool' or 'which aspects of its observed use are essential and relevant, and which are superfluous' (Csibra and Gergely 2006: 4).

These reasons for denying the sufficiency of imitation strike me as too local in several respects. First, they focus exclusively on tool use, but our hominid ancestors are likely to have had many more targets for cultural inheritance. Indeed, Csibra and Gergely's own helpful list of culturally inherited traits includes not only the 'function and use of tools', but the 'valence of objects or animals, some aspects of language (primarily words), non-linguistic symbols (for example, gestures), cultural conventions, and even abstract beliefs expressing the world view of the community' (Csibra and Gergely 2006: 11). Consequently, it is possible that any 'learnability problem' was more general: that it related to the encoding of gestures and vocalizations, as well as the construction of tools.

Second, Csibra and Gergely's reasons presume the soundness of a controversial theory of the way that children learn about action. The teleological theory is certainly interesting but it is not widely accepted, and for some alternative theories second-order tools present a quite different learnability problem. For example, there is a wealth of evidence that infants and children learn how to use tools via domain-general mechanisms of learning that are present in a wide range of vertebrate species (Gopnik et al. 2004; Klossek and Dickinson 2012). If this is correct, the advent of second-order tools presented a sequence learning challenge, rather than an opacity problem; learners needed to represent chains or hierarchies of action–outcome relationships, rather than purely binary relationships. However, since on this alternative account the long-established mechanisms had not been directed to identifying a definitive goal for any given action, it was not a problem that second-order tools made ultimate goals difficult to track. Thus, the teleological theory created the learnability problem that the theory of natural pedagogy is designed to solve.

Third, it is likely that what Csibra and Gergely call 'blind imitation' can solve the problems that they consider to be beyond its power. If a novice only ever got to see an expert using a second-order tool on one or a very few occasions, it would indeed be difficult for the novice to work out under what conditions the tool should be used, and 'which aspects of its observed use are essential and relevant, and which are superfluous' (Csibra and Gergely 2006: 252). But if the technology is important, novices are likely to have many opportunities to observe use of the second-order tool, and multiple demonstrations would allow domain-general processes of statistical learning to extract recurrent elements of the context and technique. Recurrent elements are likely to be essential and relevant, whereas variable or occasional elements are likely to be superfluous.

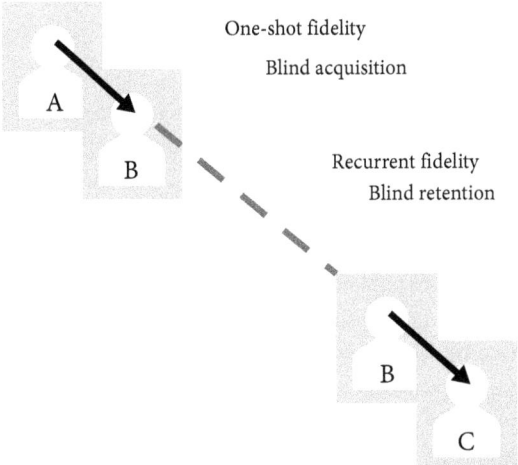

Figure 4.1 One-shot and recurrent fidelity in cultural inheritance. The solid arrows represent two episodes in the cultural inheritance of a trait, t: when t is learned by agent B from agent A (upper arrow), and later when t is learned by agent C from agent B (lower arrow). 'One-shot fidelity' refers to the extent to which t is unchanged within social learning episodes of this kind. The broken line represents the interval between these episodes of social learning; the period after t is learned by B from A and before t is learned by C from B. 'Recurrent fidelity' refers to the extent to which t is unchanged during this kind of interval. The fidelity of cultural inheritance as a whole depends on both one-shot and recurrent fidelity. Each of these types of fidelity require blind trust—i.e. processes that incline agents to adopt t (one-shot fidelity, blind acquisition) and retain t (recurrent fidelity, blind retention) regardless of what their personal experiences may imply about the instrumental function of t.

Finally, and most important in my view, Csibra and Gergely's reasons for thinking that imitation is not enough are too local in focusing, as most psychologists do, on 'one-shot fidelity': on the fidelity with which a trait, t, is initially learned from an expert, A, by a novice, B (Figure 4.1). A fair degree of fidelity at this initial stage is undoubtedly necessary for cumulative cultural evolution, but it is radically insufficient. For improvements to accumulate, 'recurrent fidelity' is needed: B must retain t—keep doing what A did, or keep believing what A believed—until C, a novice of the next cultural generation, acquires t from B. In other words, t needs to be insulated from modification by individual learning between acquisition and 're-transmission'—and, as far as I am aware, there is no evidence that acquisition by imitation, or any other kind of social learning, has this insulating effect (Heyes 1993; Shea 2009).[2]

---

[2] The results of a recent study by Hernik and Csibra (2015) suggest that infants retain information about tool functions, in the face of 'counter-evidence', more reliably when the tool functions are demonstrated with infant-directed speech than in the absence of ostensive cues. Ultimately, this line of research may reveal that learning with ostensive cues promotes recurrent fidelity. However, to find out whether the effect is specific to ostensive cues and to information retention, it will be necessary to compare ostensive cues with other attention-grabbing stimuli, and to establish that infants who see ostensive and control

## 4.4 Smart Thinking and Blind Trust

Most researchers who recognize that imitation is not enough seek the 'extra' among the fanciest—the most complex, or at least the most obscure—processes in the catalogue of cognitive science. They suggest that if one adds theory of mind (also known as 'mind reading', 'mentalizing', and 'shared intentionality') and/or language to imitation, the resulting compound is enough to support cumulative cultural inheritance (Byrne and Rapaport 2011; Tomasello 2014). But Csibra and Gergely have taken a different path, and one that is better suited to explaining the early origins of cultural inheritance: they ask how cultural evolution got off the ground (Sterelny 2012). The theory of natural pedagogy suggests that blind trust is at least as important as smart thinking. When it comes to acquiring cultural traits, infants and children need to take what they are given, regardless of whether it makes sense to them. Indeed, according to Csibra and Gergely, children should be especially inclined to learn from others the things that *don't* make sense to them.

Working from their teleological account of action learning, and focusing on one-shot fidelity, Csibra and Gergely suggest that blind trust is necessary because some cultural traits, such as those involved in mediated tool use, cannot be acquired via insightful, teleological thinking, and suggest that the blind trust that solves this problem is the kind embodied in their theory of pedagogy: a propensity on the part of novices to learn selectively what an expert directs them to learn. In the previous section, I suggested a broader perspective in which the insufficiency of imitation applies across different types of cultural trait and lies mainly in its incapacity to insulate them from individual learning for recurrent fidelity. From this broader perspective, the kind of trust embodied in pedagogy is likely to be part of the solution, but other insulation mechanisms will be at least as important. A tendency to attend to, and therefore to learn, the things that an expert intends to teach will increase the fidelity with which $B$ inherits $t$ from $A$, but the biases postulated by the theory of natural pedagogy—whether inborn or otherwise—will not, by themselves, increase the fidelity with which $B$ retains $t$ until $t$ can be inherited by $C$. Faithful cultural inheritance requires recurrent as well as one-shot fidelity, blind retention as well as blind acquisition (Figure 4.1).

Tomasello and his colleagues suggest that human children have two psychological attributes that seem well suited to promoting recurrent fidelity through blind retention, i.e. insulation of $t$ between acquisition by $B$ and re-transmission to $C$. These are 'social motivation', an intrinsic desire to be like others in one's social group, and 'normative thinking', including a capacity to think 'I must do it the way I was taught' (Schmidt et al. 2011; Tomasello 2014). The broader perspective I am offering suggests that attributes such as social motivation and normative thinking may be more important in relation to cultural inheritance than the theory of natural pedagogy

---

demonstrations have learned equally from the demonstrations before they are exposed to counter-evidence. Without the latter check, any effect on retention of ostensive cues would be reducible to an effect on acquisition, and therefore unlikely to have long-term effects.

acknowledges. However, as discussed elsewhere (Heyes 2018), I think Csibra and Gergely are rightly reluctant to treat whole, big psychological attributes—like social motivation and normative thinking—as primitives when trying to explain the origins of cultural learning.

## 4.5 Genetic Tweaking is a Powerful Source of Cognitive Change

On my intentional stance reading of the theory of natural pedagogy (see section 4.2), one of its central claims is that small changes to psychological processes—inflections or tweaks (Heyes 2003; Milius 2013) that create biases in favour of certain perceptual inputs—can make a huge difference to the way the processes function. For example, natural pedagogy theory shows us that tweaking imitation, so that it is more likely to occur when a demonstration is preceded by eye contact, and tweaking motion cuing, so that it becomes especially sensitive to eyes (i.e. gaze cuing), could transform imitation and motion-cuing into major conduits of high-fidelity cultural inheritance (Heyes 2016). I think this emphasis on the power of small changes is exactly right; but we should not assume that the small-but-crucial changes are underpinned by genetic variation—that genetic evolution is doing the tweaking.

The evidence reviewed in Heyes (2016) indicates that imitation may not be as reliably modulated by eye contact as the theory of natural pedagogy suggests, and that both imitation and motion cuing may be biased or specialized by domain-general processes of learning in the course of development, rather than having been tweaked by genetic evolution in the hominid line. But this evidence does not undermine Csibra and Gergely's basic approach, their attempt to find the roots of cultural inheritance in small changes to psychological processes. Rather, it suggests that we should be more catholic when examining the drivers of change, more open to the possibility that the crucial changes are made not by genetic evolution, but by domain-general processes of learning operating in a sociocultural context.

This openness could, I think, be usefully extended from the components of natural pedagogy to the components of 'shared intentionality' (Tomasello 2014), including social motivation and normative thinking, which have the potential to support recurrent fidelity. Social motivation is a good example because, in the last decade, the idea that human infants and children find it rewarding to behave in the same way as those around them—that they are socially motivated—has been revived by evidence of 'overimitation' (Over and Carpenter 2013). Studies of overimitation suggest that children copy the behaviour of adults even when, from an instrumental perspective, it is inefficient to do so. For example, when retrieving a toy from a puzzle box, three- and four-year-old children do not only release the latches and open the doors impeding their access to the prize. They also copy the model's extraneous actions—such as tapping the box with a wand—and engage in this overimitation even when they are apparently able to discriminate the 'silly', extraneous actions from the actions necessary to get the job done (Lyons et al. 2007).

Whether or not they call it social motivation, contemporary researchers typically assume that the impulse to overimitate is a genetic adaptation or 'evolved heuristic' (Whiten et al. 2009)—that human children have an inborn tendency or pre-disposition to overimitate. But evidence published predominantly in the 1970s and 1980s—when overimitation was called 'generalized imitation'—suggests that this assumption is not sound. In these studies, children from pre-school age up to 14 years were tested for imitation of a set of novel actions at baseline—i.e. before imitation of any action in the set had been rewarded—and after selective reward—i.e. when imitation of some, but not all, actions in the set had begun to be rewarded. The frequency of imitation increased substantially from baseline when the reward, typically praise, was introduced (Baer and Sherman 1964), and the children showed generalization (overimitation): once the reward had been introduced, the children were apt to imitate not only the actions for which imitation was rewarded, but also other actions of the same type (Baer and Sherman 1964; Garcia et al. 1971; Young et al. 1994).

At the time, these findings were interpreted within a behaviourist framework (Baer and Deguchi 1985), but one does not need to subscribe to any particular theory of learning to feel the impact of these studies. They show that experiences of social reward are a powerful determinant of imitation in childhood, and thereby encourage research that investigates rather than assumes that the impulse to overimitate is inborn. One alternative possibility is that in humans, genetic evolution has simply increased the reward value of response-contingent stimulation. Since imitation delivers response-contingent stimulation (what the imitator does depends on what the imitatee has just done), this would produce a positive feedback loop: Being imitated by a child (or anyone else) makes an adult feel good, and consequently likely to react in a positive way—to move towards the child, smile, or say something nice (Grusec and Abramovitch 1982). The child finds the adult's reaction rewarding—partly because the reaction was positive, but also just because it was response-contingent—making the child more likely to imitate the same and similar behaviour now and in the future. It doesn't matter whether the behaviour is 'silly' in the sense that it plays no role in securing an asocial reward. Research on generalized imitation suggests that as long as imitation of similar behaviour has secured social rewards in the past, the impulse to 'overimitate' will be there.

This alternative hypothesis suggests that genetic evolution has tweaked rather than reconstructed social motivation in the hominid line. Rather than giving us a specific new desire—to act like others in our social group—Mother Nature may have simply amplified the rewarding power of response-contingent stimulation.

In summary: I have suggested that the theory of natural pedagogy embodies three insights that it would be wise to pursue in future research on cultural learning: imitation is not enough; the extra comes from blind trust; and tweaks are a powerful source of cognitive change. In discussing each of these insights, I have argued that imitation is not enough primarily because it does not, by itself, promote recurrent fidelity; that the extra comes from processes, such as social motivation and normative thinking, that promote blind retention, as well as from processes that promote blind acquisition; and that genetic tweaks are a powerful source of cognitive change because their effects can be amplified massively by domain-general processes of learning operating in a sociocultural context (Heyes 2018).

## 4.6 Evolutionary Causal Essentialism

As far as I am aware, Csibra and Gergely have never said that natural pedagogy is part of 'human nature'. However, their inclusion of 'natural' in the name of the theory, defence of the view that pedagogy occurs in all cultures (Csibra and Gergely 2011), and emphasis on the evolutionary origins of natural pedagogy make natural pedagogy an interesting test case for theories of human nature. In particular, it seems that two recent accounts of human nature—Machery's (Chapter 1 this volume) nomological account and Samuels's (2012) causal essentialist theory—should embrace natural pedagogy. According to both Machery and Samuels, an important function of a theory of human nature is to explicate the practices of cognitive scientists engaged in just the kind of project represented by the theory of natural pedagogy. Therefore, the nomological account and the causal essentialist theory should each characterize human nature such that if the claims of the theory of natural pedagogy were true, natural pedagogy would be a component of human nature.

In practice, both accounts capture natural pedagogy, but not very firmly. The nomological account says: 'Human nature is the set of properties that humans tend to possess as a result of the evolution of their species', combining 'the universality proposal'—that traits belonging to human nature must be typical of human beings—and 'the evolution proposal'—that they must be evolved traits (Machery, Chapter 1 this volume). This certainly encompasses Csibra and Gergely's claims that natural pedagogy is found in all cultures, and that it is a set of specific genetic adaptations. However, because the nomological account allows that *all* traits we possess as a result of the evolution of our species are part of human nature—e.g. morphological, physiological, behavioural, and cognitive traits—Machery's account does not have a firm grasp on Csibra and Gergely's preoccupation with generative mechanisms. Like much of evolutionary psychology, broadly construed (Heyes 2003), the theory of natural pedagogy is concerned not so much with the manifest characteristics of our species—with how adult humans could be distinguished from other species by a newly arrived Martian—but with the underlying mechanisms that make us different; specifically, with the ways in which we learn from others in infancy and childhood.

On the causal essentialist view, 'human nature is a suite of mechanisms that underlie the manifestation of species-typical cognitive and behavioural regularities'. Thus, in contrast with the nomological account, the causal essentialist view captures exactly Csibra and Gergely's concern with generative mechanisms. It stipulates that components of human nature are 'species-typical', and is therefore also consistent with the claim that natural pedagogy occurs in all cultures. However, the causal essentialist account does not confine human nature to mechanisms that have particular origins, and so does not make sense of Csibra and Gergely's dominant concern with the evolutionary origins of natural pedagogy.

Here is a simple suggestion: why not cross the nomological account with causal essentialism? This hybrid would make human nature the set of mechanisms that underlie the manifestation of species-typical cognitive and behavioural regularities, which humans tend to possess as a result of the evolution of their species. In addition to providing a snug fit for natural pedagogy, this hybrid account, which we can call 'evolutionary causal essentialism', inherits some significant assets from its parents. Like the nomological view, it is evolutionary, but it does not commit the essentialist sins found by evolutionists in older theories of human nature (Hull 1986). For example, it does not offer necessary and sufficient conditions for being human, nor imply that human nature is either normative or fixed. Similarly, evolutionary causal essentialism, like causal essentialism, preserves the causal-explanatory function of human nature; following Aristotle, Locke, and Hume, it casts the components of human nature as underlying or 'hidden' entities that explain obvious differences between humans and other creatures (Samuels 2012). Evolutionary causal essentialism may also escape some of the problems encountered by its parents. For example, the nomological view has been accused of excessive liberalism, because it allows that Sikhism and skiing are components of human nature (Lewens 2015). In contrast, evolutionary causal essentialism, like plain causal essentialism, would not admit Sikhism and skiing, however widespread they became, because they are sets of manifest characteristics (behaviours, beliefs, values), rather than generative mechanisms of the kind studied by cognitive science.

A more serious charge against the nomological view is that it makes the arbitrary assumption that a characteristic can be a component of human nature if it is a product of the genetic evolution of our species, but not if it results from social learning (Lewens 2015) or cultural evolution (in this volume, Laland and Brown, Chapter 7, and Richerson, Chapter 8). To overcome this problem, and on independent grounds (e.g. Heyes 2012, 2016, 2018; Heyes and Frith 2014), I favour a version of evolutionary causal essentialism in which 'the evolution of our species' is understood to encompass not only genetic processes but cultural and epigenetic inheritance systems.[3] On this reading, evolutionary causal essentialism would have a firm hold on natural pedagogy—it would fulfil the rationalizing and scientific project-defining functions of a theory of human nature (in this volume, Kronfeldner, Chapter 10, Machery, Chapter 1, and Sterelny, Chapter 6;

---

[3] My instinct (pun intended) would be to count as contributing to 'the evolution of our species' only processes that are based on variation-and-selective-retention (Campbell 1974). Whatever the merits of that particular approach, evolutionary causal essentialism certainly needs some well-founded constraints on what it takes to be an evolutionary process.

Samuels 2012)—even if Mother Culture plays a more significant role than Mother Nature in preparing humans to be taught.

## 4.7 Conclusion

The theory of natural pedagogy proposes that Mother Nature has played a major role in making human infants and children receptive to teaching signals. With this specific 'purpose', genetic evolution has introduced at early stages in development small but crucial changes in the way that human infants process information from other people. The evidence reviewed in Heyes (2016) and summarized at the beginning of this chapter, combined with the arguments presented here, suggest that Mother Nature may have had less specific purposes when she did her tweaking, and that Mother Culture does a lot of the hard work in preparing children to be taught.

Less specific purposes are suggested by the fact that where there is evidence that a component of natural pedagogy has been shaped by genetic evolution, there is also evidence that it was adapted not for teaching, but for social bonding or to promote attention to other agents. A more important role for Mother Culture is suggested by evidence that some components of the pedagogy package (infant-directed speech, gaze cuing, rational imitation), and other psychological attributes (social motivation, overimitation), become adapted for teaching through the operation of domain-general processes of learning in social contexts. For example, the evidence suggests that motion cuing becomes gaze cuing, and possibly modulated by communicative cues, through reinforcement learning in contexts where the eye movements of others predict the locations of interesting objects and events. Similarly, research on generalized imitation suggests that children may become overimitators through reinforcement learning in contexts where adults deliver rewards for imitation because they, the adults, judge the behaviour to be similar to their own, or simply because, without recognizing that it is imitative, the adults are pleased by the response-contingent character of the children's imitative behaviour.

In these examples, learning that makes a child more teachable is guided by the actions of adults. If the adults' guiding actions, such as gaze shifting and rewarding imitative behaviour, are intended to support learning by the child, then children are taught to be teachable. But even when adults do not intend to influence a child—when they are going about their normal business, looking at events that interest them, and reacting warmly to behaviour simply because they find it pleasing—the effect of their actions is to promote the development of psychological tendencies that make children teachable; natural pedagogy, or 'teachability', is culturally inherited. According to the nomological account, this disqualifies natural pedagogy from the realm of human nature. In contrast, evolutionary causal essentialism recognizes cultural inheritance as part of 'the evolution of our species', and therefore admits natural pedagogy to the warm embrace of human nature.

## References

Baer, D. M., and Deguchi, H. (1985). 'Generalized Imitation from a Radical-Behavioral Viewpoint.' In S. Reiss and R. Bootzin (eds), *Theoretical Issues in Behavior Therapy*, 179–217. Orlando, Fla.: Academic Press.

Baer, D. M., and Sherman, J. A. (1964). 'Reinforcement Control of Generalized Imitation in Young Children.' *Journal of Experimental Child Psychology* 1: 37–49.

Beisert, M., Zmyj, N., Liepelt, R., Jung, F., Prinz, W., and Daum, M. M. (2012). 'Rethinking "Rational Imitation" in 14-Month-Old Infants: A Perceptual Distraction Approach.' *PloS One* 7: e32563.

Boyd, R., and Richerson, P. J. (1988). *Culture and the Evolutionary Process*. Chicago: University of Chicago Press.

Bruner, J. (1983). *Child's Talk*. Oxford: Oxford University Press.

Byrne, R. W., and Rapaport, L. G. (2011). 'What Are We Learning from Teaching?' *Animal Behaviour* 82: 1207–11.

Campbell, D. T. (1974). 'Evolutionary Epistemology.' In P. A. Schlipp (ed.), *The Philosophy of Karl Popper*, 413–63. LaSalle, Ill.: Open Court.

Cavalli-Sforza, L. L., Feldman, M. W., Chen, K.-H., and Dornbusch, S. M. (1982). 'Theory and Observation in Cultural Transmission.' *Science* 218: 19–27.

Csibra, G., and Gergely, G. (2006). 'Social Learning and Social Cognition: The Case for Pedagogy.' *Processes of Change in Brain and Cognitive Development: Attention and Performance XXI*, 21: 249–74.

Csibra, G., and Gergely, G. (2011). 'Natural Pedagogy as Evolutionary Adaptation.' *Philosophical Transactions of the Royal Society B* 366: 1149–57.

Custance, D. M., Whiten, A., and Bard, K. A. (1995). 'Can Young Chimpanzees (*Pan troglodytes*) Imitate Arbitrary Actions? Hayes and Hayes (1952) Revisited.' *Behaviour* 132: 837–59.

Dennett, D. (2009). 'Intentional Systems Theory.' In B. McLaughlin, A. Beckerman, and S. Walter (eds), *The Oxford Handbook of the Philosophy of Mind*, 339–50. New York: Oxford University Press.

Garcia, E., Baer, D. M., and Firestone, I. (1971). 'The Development of Generalized Imitation within Topographically Determined Boundaries.' *Journal of Applied Behavior Analysis* 4: 101–12.

Gergely, G., and Csibra, G. (2003). 'Teleological Reasoning in Infancy: The Naïve Theory of Rational Action.' *Trends in Cognitive Sciences* 7: 287–92.

Gergely, G., and Csibra, G. (2005). 'The Social Construction of the Cultural Mind: Imitative Learning as a Mechanism of Human Pedagogy.' *Interaction Studies* 6: 463–81.

Gergely, G., Nádasdy, Z., Csibra, G., and Biro, S. (1995). 'Taking the Intentional Stance at 12 Months of Age.' *Cognition* 56: 165–93.

Gopnik, A., Glymour, C., Sobel, D. M., Schulz, L. E., Kushnir, T.. and Danks, D. (2004). 'A Theory of Causal Learning in Children: Causal Maps and Bayes Nets.' *Psychological Review* 111(1): 3–32.

Grusec, J. E., and Abramovitch, R. (1982). 'Imitation of Peers and Adults in a Natural Setting: A Functional Analysis.' *Child Development* 3(3): 636–42.

Hernik, M., and Csibra, G. (2015). 'Infants Learn Enduring Functions of Novel Tools from Action Demonstrations.' *Journal of Experimental Child Psychology* 130: 176–92.

Heyes, C. (1993). 'Imitation, Culture, and Cognition.' *Animal Behaviour* 46(5): 999–1010.
Heyes, C. (2003). 'Four Routes of Cognitive Evolution.' *Psychological Review* 110(4): 713–27.
Heyes, C. (2012). 'Grist and Mills: On the Cultural Origins of Cultural Learning.' *Philosophical Transactions of the Royal Society B* 367: 2181–91.
Heyes, C. (2016). 'Born Pupils? Natural Pedagogy and Cultural Pedagogy.' *Perspectives on Psychological Science* 11: 280–95.
Heyes, C. (2018). *Cognitive Gadgets: The Cultural Evolution of Thinking*. Cambridge, Mass.: Harvard University Press.
Heyes, C., and Frith, C. D. (2014). 'The Cultural Evolution of Mind Reading.' *Science* 344. doi: 10.1126/science.1243091.
Hull, D. L. (1986). 'On Human Nature.' *Proceedings of the Biennial Meeting of the Philosophy of Science Association* 2: 3–13.
Klossek, U. M. H., and Dickinson, A. (2012). 'Rational Action Selection in 1½-to 3-year-Olds Following an Extended Training Experience.' *Journal of Experimental Child Psychology* 111: 197–211.
Lewens, T. (2015). *Cultural Evolution*. Oxford: Oxford University Press.
Lyons, D. E., Young, A. G., and Keil, F. C. (2007). 'The Hidden Structure of Overimitation.' *Proceedings of the National Academy of Sciences* 104(50): 19751–6.
Milius, S. (2013). 'A Different Kind of Smart: Animals' Cognitive Shortcomings Are as Revealing as Their Genius.' *Science News* 183: 24–9.
Nakao, H., and Andrews, K. (2014). 'Ready to Teach or Ready to Learn: A Critique of the Natural Pedagogy Theory.' *Review of Philosophy and Psychology* 5: 465–83.
Over, H., and Carpenter, M. (2013). 'The Social Side of Imitation.' *Child Development Perspectives* 7: 6–11.
Piaget, J. (1962). *Play, Dreams, and Imitation in Children*. New York: Norton.
Samuels, R. (2012). 'Science and Human Nature.' *Royal Institute of Philosophy Supplement* 70: 1–28.
Schmidt, M. F. H., Rakoczy, H., and Tomasello, M. (2011). 'Young Children Attribute Normativity to Novel Actions without Pedagogy or Normative Language.' *Developmental Science* 14: 530–39.
Shea, N. (2009). 'Imitation as an Inheritance System.' *Philosophical Transactions of the Royal Society B* 364: 2429–43.
Sterelny, K. (2012). *The Evolved Apprentice*. Cambridge, Mass.: MIT Press.
Tomasello, M. (2014). *A Natural History of Human Thinking*. Cambridge, Mass.: Harvard University Press.
Washburn, M. F. (1908). *The Animal Mind*. New York: Macmillan.
Whiten, A., McGuigan, N., Marshall-Pescini, S., and Hopper, L. M. (2009). 'Emulation, Imitation, Over-Imitation, and the Scope of Culture for Child and Chimpanzee.' *Philosophical Transactions of the Royal Society B* 364: 2417–28.
Young, J. M., Krantz, P. J., McClannahan, L. E., and Poulson, C. L. (1994). 'Generalized Imitation and Response Class Formation in Children with Autism.' *Journal of Applied Behavior Analysis* 27: 685–97.

# 5

# Human Nature
## A Process Perspective

*John Dupré*

## 5.1 Introduction

According to the ever-reliable Wikipedia, 'Human nature refers to the distinguishing characteristics—including ways of thinking, feeling and acting—which humans tend to have naturally, independently of the influence of culture.'\* This simple statement summarizes several of the central features of many mainstream views on this topic.

First is the proposal that human nature is a set of distinguishing characteristics. In its strongest form, this is the view that human nature is the human essence, a necessary and sufficient condition for something to be human. But weaker interpretations—for example, clusters of properties some sufficient number of which will qualify an entity as human—or even defeasible Wittgensteinian criteria for humanness, might suffice.

A second noteworthy point is the emphasis on the psychological. This is, indeed, what people almost always have in mind when talking about human nature, and this is further foregrounded by the following contrast with culture. I shall follow this tradition in much of what follows, but it is worth pointing out that this deviates quite significantly from more general uses of the term 'nature' with reference to the living world. With reference to complex animals, we might emphasize behaviour or action as central to their nature, but hardly thought or feeling. For plants or bacteria, we are likely to think of physiology or morphology.

Finally, there is the contrast with culture. This is very often exactly the point of appeals to human nature. If someone does something that is characteristic of human nature, it is otiose to appeal to more specific social, personal, or cultural explanations. This is just what people do, regardless of any such contingencies.

---

\* I am very grateful to Tim Lewens for comments on an earlier draft that led to a number of significant improvements. The research leading to this chapter has received funding from the European Research Council under the European Union's Seventh Framework Programme (FP7/2007-2013)/ERC grant agreement n° 324186.

I think all these aspects of this simple definition are problematic. I shall begin by focusing on the first and apparently most innocuous. What can be wrong with distinguishing the (typical) characteristics of humans?[1] And if that's all that the concept of human nature means, how can it do any harm? A familiar response to these questions is that describing a set of characteristics as 'human nature' tends to distract attention from the diversity of humans and, worse, to make humans who lack such characteristics appear as abnormal or unnatural. I think this is a legitimate concern, but here I want to start in a slightly different place.

Supposing I say that it is part of human nature to speak a language. It is true that not everyone can speak a language, as people can lack some of the many physiological attributes that underpin this capacity. In view of the existence of deaf people, many of whom have a language but do not *speak* it, we might do better to write of using a language than of speaking it. For those who lack any language capacity, it is perhaps not controversial to attribute to them a serious deficiency, a lack of a normal part of human nature. Any reasonable concept of a biological nature will allow such exceptions.

But there is a more serious difficulty. Humans a few months old, or human foetuses, do not have language. Some people, in extreme old age, lose some or all of their ability to speak. This simple point is in fact of great importance. If there is a human nature, it does not consist just of a set of properties that humans possess, but of properties that humans possess, or typically possess, at particular stages of their lives. A human is not a thing with a fixed set of properties, but a life cycle. I suggest that, as such, we are better thought of as processes than as things or, in traditional philosophical language, substances.

The distinction between things and processes is an ancient one. The thesis of the ontological centrality of processes is generally traced back to Heraclitus, but his view on the whole lost out to the thing-centred atomism of Democritus, Parmenides, and others. While a number of subsequent philosophers have been sympathetic to a process ontology, again the majority are more or less loosely aligned with the substance view.[2] The growing influence of science, and of the atomism of leading contributors to the scientific revolution (Locke, Boyle, and so on), did much to embed this dominance.[3]

The well-known exception in the last century has been Alfred North Whitehead,[4] whose process philosophy has attracted a number of philosophical followers as well as—perhaps equally significantly for present purposes—the attention of several prominent theoretical biologists, including Conrad Waddington, Joseph Needham, and Ludwig von Bertalanffy. Though Whitehead's work undoubtedly encompasses crucial insights, I shall not discuss it here. First, Whitehead uses a good deal of esoteric

---

[1] A prominent philosophical defender of the concept of human nature, Edouard Machery (e.g. 2008, and Ch. 1 this volume) follows this kind of line, requiring only that ingredients of human nature be sufficiently widespread, and widespread as a result of evolutionary processes.
[2] For a clear introduction to process philosophy, see Rescher (1996).
[3] Though it is perhaps surprising that the rise of quantum mechanics did not reverse this tendency.
[4] The *locus classicus* is Whitehead (1978).

terminology that many have found off-putting. And second, his philosophical system also includes theological and idealistic elements that I wish to avoid. I find it easier to risk reinventing the wheel than to articulate my differences from Whitehead in detail. But in common with Whitehead and other process philosophers, I believe that the assumption of a substance ontology has been a disaster for our thinking about human nature.

## 5.2 Humans as Processes

A process necessarily persists for some length of time. Within their temporal extension, many processes have an origin, sequence of states, and conclusion that are more or less similar to the corresponding stages of a larger class of processes, and this fact makes possible the allocation of processes to kinds, such as thunderstorms and battles. More important examples for present purpose are living systems. The human process is naturally identified with the typical human life cycle—typical, though of course life cycles are often cut short, and they may deviate from the norm in countless ways.

As I have noted, properties characteristic of humans will vary according to the stage of the life cycle. On the other hand, this variation is far from random. Human life cycles follow characteristic patterns. So, provided we index properties with stages of the life cycle, does it make much difference moving to the process perspective? To address this question, we may begin by attending to a further important difference between a substance and a process. The default condition for a substance is stasis: if it has a particular set of properties at some time, other things being equal it will continue to do so at later times. Explanation of some kind is called for when its properties change. For a process, on the other hand, its very persistence will normally require explanation. For a thunderstorm, say, to persist, the matter that composes it must continue to move in specific ways if it is not just to cease or to dissipate. Properties of the constituent matter and features of the whole system combine to explain why the relevant movement persists. Arguably, the most important difference between a substance and a process is that whereas for the former the central problems, both empirically and philosophically, concern the explanation of change, for the latter the central problems concern understanding stability, or persistence.

Much of biology, in fact, is engaged in explaining the stability of biological systems. To understand this point properly, it is necessary to appreciate the hierarchical nature of biological systems—not so much (as is commonly considered) the spatial hierarchy, but rather the temporal one. Cell biologists describe sequences of events that contribute either to the persistence of the cell (e.g. RNA transcription or oxidative phosphorylation) or to the execution of the cell cycle (e.g. the various processes that constitute mitosis). While these are sequences of changes, they are also central elements of the explanation of the stability of the cell or the multicellular organism of which they are part. And indeed, on an evolutionary timescale, it is this contribution that explains their occurrence. The cell cycle, as well as serving the short-term persistence of the organism, in conjunction with processes of cell differentiation and death (apoptosis) is at the heart of the explanation of a cyclical process at a much longer timescale, namely, development. Finally, developmental cycles explain, in part, the stability of the lineages that constitute the evolutionary process.

One reason why it is so important to keep in mind these differences of temporal scale is that they allow us to explain something that a process perspective certainly had better come to terms with: the success of mechanistic explanations. Mechanism, a philosophical view currently undergoing a major resurgence, is a quintessentially substantialist position (Machamer et al. 2000; Glennan 1996; for criticism, see Nicholson 2012, 2013.). Mechanism prescribes the explanation of the behaviour of complex systems such as are addressed by the life sciences in terms of the properties and interactions of their parts. As a matter of fact, I am somewhat suspicious of the very idea of a unique and objective decomposition of biological systems into their parts.[5] Nonetheless, there are, typically, entities that are stable on the timescale of a particular process we are trying to understand. When, for example, we explain metabolic processes, it is often the case that important enzymes or cell membranes can be treated as stable things. But of course these may also be understood as stages in processes of construction and degradation. This perspective on mechanistic explanation should not only account for its considerable successes but also point to its limits, the places where it can be expected to break down.

This kind of stabilization also applies to organisms. At much shorter timescales than the transitions between life cycle stages, humans undergo a wide variety of processes—for example, metabolic—that maintain their short- or medium-term stability. Basic processes such as oxidative phosphorylation occur at all stages of the life cycle, though perhaps less efficiently at very late stages. Again, it is essential to keep in mind a hierarchy of processes. An organism lies at the intersection of a very large number of subordinate processes that together maintain its short-term stability and longer-term developmental trajectory.

The stable cycle of stages through which human lives typically pass may be viewed from a different perspective, as involved in the stabilization of a process of much longer timescale, the human lineage. This is also one way of understanding the connection of human behaviour with evolution, though, as I shall suggest below, one that carries rather different implications from those of more familiar accounts of this relation. For now I note only the obvious points that human life cycles are the constituents of the human lineage, and the success of the former in persisting and launching new such cycles is the main determinant of the persistence of the latter.

As well as processes that pass through a sequence of stages, like human life cycles, many, like storms or eddies, have no particular trajectory. They may nonetheless be very stable. The Red Spot on Jupiter, a gigantic storm on the planet's surface, has persisted

---

[5] It is no surprise that the approach to human nature that gives least respect to the dynamic features of human development and human evolution, evolutionary psychology, sees the mind as divided into a quite specific and largely autonomous set of parts or modules.

for centuries, albeit with some changes in shape and location. This persistence is not just a default, in the absence of any intervention, but requires a lot of activity—in the case of the Red Spot high-speed prevailing winds that maintain its boundaries and provide the energy of its rotating matter. Were these sustaining processes to stop, the Red Spot would instantly disintegrate. If, as process philosophers generally suppose, at the most fundamental level reality is flux, then some explanation is needed wherever we encounter persistent structure. Whereas mechanists are concerned only with explaining how structures enable the execution of functions, process philosophers see also the necessity of explaining how function maintains structure.

## 5.3 Process and Plasticity

Since living things exist in complex and changing environments, the best way to persist is not always determinable in advance; it will often be adaptive, to have the capacity to respond flexibly to environmental contingencies. Successful living systems, in fact, display plasticity at multiple levels. The extent of this plasticity, right down to the molecular level, has been one of the most striking themes of the biology of the last few decades. However, and as already noted, discussions of human nature are generally directed especially to characteristic human behaviour, and from now on this will be the main topic here; certainly, nowhere are the benefits of plasticity more important than in the realm of behaviour.

It is tempting to suggest that distinctive modes of flexibility characterize the major different modes of organization of life. For microbes, plasticity is chemical or metabolic. Microbes appear to be able to extract energy from almost any chemical substance from which this is in principle possible. They do this, frequently, by forming consortia of different microbes providing different chemical resources. Some of these resources are packaged in forms, such as in plasmids, that enable them to be shuffled conveniently from one microbial cell to another. These chemical capacities of microbes are exploited by all or most multicellular eukaryotic organisms, most obviously (though by no means exclusively) in digestion. Humans, in particular, exist in symbiotic relations with trillions of microbes, which contribute crucially to processes including digestion, immune response, development, and more. These symbioses provide one dimension of plasticity, or responsiveness to changes in the environment, at a far shorter timescale than that envisaged in standard models of evolutionary adaptation.

Plants, while outsourcing many of their metabolic needs either to independent microbial consortia or to long-captive endosymbionts (plastids), have introduced a new form of plasticity in morphology. The growing tip of a plant process, or meristem, includes a collection of stem (*sic*) cells that allow differentiation into various organs (leaves, branches, flowers) in response to the affordances offered by the environment. As gardeners who engage in such practices as taking cuttings or layering will be aware, the extent of this plasticity is highly variable; but all plants exhibit such morphological plasticity to some degree. The different capacities of the meristem also allow rapid morphological evolution, as strikingly demonstrated in the artificial differentiation of the wild cabbage *Brassica oleracea* into such morphologically diverse forms as cabbage, broccoli, cauliflower, Brussels sprouts, and kohlrabi.

Multicellular animals, on the other hand, generally adopt relatively determinate morphologies. I should emphasize the word 'relatively'. As is demonstrated in Mary Jane West-Eberhard's (2003) encyclopedic study of developmental plasticity, animal morphology is by no means fully fixed. What is clear, at any rate, is that animals have added a whole new range of flexible response to the environment, through evolving nervous systems that allow the production of behaviour responsive to environmental contingencies.

It is useful to distinguish several grades of behavioural flexibility. At the very simplest, least plastic level are little more than reflexes, typically automatic responses to potentially harmful or beneficial stimuli, such as the chemotaxis found in many bacteria. Much more complex are the inflexible sequences of behaviour found, for instance, in many arthropods. Examples well known to philosophers are the provisioning behaviours of sphecid wasps discussed in several places by Daniel Dennett, or the nest-building behaviour of leaf-cutter bees of the genus *Megachile*. Despite their impressive complexity, simple experiments appear to demonstrate that these behaviours are produced in inflexible sequences that cannot respond appropriately to changes in the animal's circumstances.[6]

Behaviour that is composed of stereotypical responses to specific environmental cues may characterize a good deal of arthropod behaviour. But if so, such behaviour need be in no way simple. Think, for instance, of the well-known waggle dance of the honeybee. This quite elaborate performance is elicited by the discovery of a pollen source, and produces in a suitable audience a specific response: movement to the pollen source. Such behaviour becomes particularly impressive when it is integrated at the social level, as the waggle dance appears to be involved in controlling the distribution of pollen among members of the colony to maintain an optimal food supply. More striking still is the organization of stereotypic behaviour in termites to maintain the structure of a mound that, among other things, provides a constant near-optimal temperature. This is an instance of niche construction, to which I shall return below.

The preceding examples exhibit activities of the organism that modulate its interaction with the environment in ways that are beneficial to its survival and reproduction. They do not emphasize elements that are characteristic of the behaviour of 'higher' animals and that are carried to unprecedented levels in humans, namely, learning and intelligence. These are both, needless to say, enormously difficult and contentious topics. They are also, however, at the core of human nature, and of why a description of human nature in the standard sense is unattainable. The importance of learning is that

---

[6] Actually, while this inflexibility has been widely asserted by philosophers and cognitive scientists, a careful study of the history of research on sphecid wasps by Fred Keijzer (2013) suggests that the story is really a lot more complicated, and that wasp behaviour shows a subtle mix of rigid and flexible aspects. For discussion, see Lewens (2015a: 233–7).

it implies that behaviour of the organism at later stages in its life cycle cannot be predicted without some knowledge of the series of environments that it has encountered during earlier stages. One interpretation of the importance of intelligence is that it implies that even with this knowledge, the response of the organism to an environmentally posed problem cannot be reliably predicted. Intelligence, that is to say, involves ingredients such as insight or imagination; its conclusions may be emulated by a similar or superior intelligence, but cannot be inferred from fixed features of the agent.

Seeing human behaviour in this way, as a uniquely developed capacity for flexible response to the environment, makes it unsurprising that it is difficult to provide a definitive account of human behaviour (and hence of the human nature that is supposed to govern behaviour) and unsurprising that human populations in different environments exhibit significantly different behaviour. But there is a further point. Humans constantly transform their environments to meet their needs: they build cities, cultivate fields, dam rivers. And they construct institutions, such as hospitals and schools, to provide enriched environments for the development of their offspring. Applying intelligence to these forms of behaviour, improvements are often made to this constructed niche, and the environment in which humans must survive is changed by their own behaviour. The relation between human behaviour, on the one hand, and on the other hand the niche that humans construct in which to behave is, in short, dialectical.

We now have the basic ingredients to summarize an account of human nature. Humans are life cycles during which a great deal of learning—mainly socially mediated (training) but also individually acquired—provides a changing and partly variable range of knowledge and behavioural capacities. Moreover, a developing capacity to first analyse a problem situation in the light of this knowledge and experience and then to decide on an appropriate course of action means that even adding a full knowledge of the individual's experience to a full account of the individual's original biological endowment will not enable a deterministic prediction of behaviour. Human intelligence is both fallible and creative. Behaviour may not be optimal; and it may not lie within any predetermined range of options. Moreover, as just discussed, the environment in which humans behave is constantly created and recreated by their own actions on it. Humans are not blank slates; they have a particular biological inheritance. But neither are they machines with a predetermined programme. A crucial part of their inheritance is their capacity for learning and intelligence.

## 5.4 Individual Life Cycles and Lineages

How does this account of the human fit into the popular, if not notorious, project of relating human nature to evolution? I have said nothing, so far, about natural selection. Following John Reiss's (2009) excellent book *Not by Design: Retiring Darwin's Watchmaker*, I shall not start there, but rather with the concept familiar from Darwin, but more closely connected to Cuvier, of *conditions for existence*. It would be impossible to summarize in any detail Reiss's complex and challenging argument, but at its core is the simple point that all organisms must, of logical necessity, have the properties that enable them to survive in the conditions they encounter—what he describes as satisfying the conditions for existence.[7] The stabilization I talked about in the first sections of this chapter is very close to the satisfaction of the conditions for existence that Reiss discusses. Whichever way it is described, this is a stringent condition. For a lineage—which I take to be a process on a longer timescale—to persist, not only must individual life cycles satisfy these conditions, but they must also spin off new life cycles, namely, reproduce. Continuing to meet the condition for existence of a lineage requires meeting this additional condition.

I have said that the persistence of a process requires explanation, and much of biology, from the study of metabolism to the study of behaviour, is concerned with providing such explanations. This range of investigations describes the multiple, interacting processes that enable the human life cycle to complete its course and, on the way, to reproduce itself. The latter, as just noted, contributes to the explanation of the persistence of a lineage. There is, however, a further crucial contributor to this persistence: stabilizing selection. What this means is just that the individuals who make up the lineage produce enough new individuals for that lineage to persist, despite the production of a certain proportion that—for reasons ranging from bad luck (struck by lightning, ambushed by a predator, and so on) to all manner of developmental failures—do not survive. The production of offspring must, for the lineage to persist, be sufficient to compensate for these losses. How much excess this requires will vary massively. For many fish, the chances of getting eaten in early youth are so high that millions of offspring must be produced. For modern humans in developed nations, the chances of viability are so high that little more than two offspring per female are required to sustain the population.

Aficionados of evolutionary theory will immediately suspect that this talk of stabilizing (or purifying) selection reveals an enthusiasm for the neutral theory of evolution, and this would be correct. In fact, given Reiss's denial of the gap between existence and adaptedness, standard ways of distinguishing between changes due to selection and changes due to drift become inapplicable. Neither, at any rate, makes the population better adapted. A great deal of the power of the concept of natural selection can be seen in its ability to answer (in part at least) the question of how a process, such as a lineage, persists for hundreds of millions of years. Bearing in mind that natural selection ensures that the form that the lineage takes must satisfy the conditions for existence, it is entirely plausible that what strike us as adaptations will nonetheless

---

[7] Reiss develops the argument by pointing out what follows from this simple point, namely, that contrary to Darwin and many of his successors, there cannot be a gap between the organism and its adaptedness for natural selection to bridge. The natural selection that Darwin and Wallace (importantly) contributed to our understanding is what Reiss calls 'narrow-sense selection': differential survival and reproduction among classes to the extent that this is caused by the distinguishing characteristics of these classes. But this provides a relatively small part of the explanation of the characteristics of a species. For a more detailed synopsis of Reiss's book, see Dupré (2010).

be the outcome of no more than a random walk through the rich but constrained space of biological possibility.

I do not want to disagree with what Motoo Kimura, the founder of the neutral theory, made clear: that the theory is not supposed to rule out the generation of novelties through mutation and natural selection. I only claim that this is a much rarer phenomenon than genetic drift. I also want to suggest that he might have gone further by allowing that phenotypic change, as much as genetic change, might be sufficiently explained as a chance process. However, in the case of the topic at hand—namely, human nature—the important point is rather different: human behaviour is highly plastic, substantially unconstrained by human biology. It is not that changes in human behaviour come about by chance, but rather that they are open to a variety of other explanations, ranging from broad cultural forces to individual decision-making. And such explanations can also account for the adaptedness of such traits.

In earlier work critical of evolutionary psychology (Dupré 2001, 2008), I have emphasized the importance of timescales, and the role of the indefensible assumption that at some important level of description (e.g. psychological modules), human behaviour can change only at the rate of natural selection of randomly generated alleles. One of the advantages of a process ontology is that it immediately draws attention to the importance of processes at different timescales. The persistence of lineages is facilitated by processes at a range of very different timescales. Even if, as is supposed by some enthusiasts for the Modern Synthesis, selection of genetic mutations is the fundamental process that adapts the lineage to a changing world, there are certainly much more rapid processes that enable much more agile adaptation.

To begin with, it is well known now that the same genome (in terms of its sequence) can be used in very different ways, depending on the different expression of its constituent genes. This is obvious from the diversity of cell types with sequentially identical genomes in multicellular organisms, and is now studied extensively under the rubric of epigenetics. It remains somewhat controversial whether epigenetic changes to an organism can be inherited across generations, and hence whether they can be part of an evolutionary process. I say 'somewhat controversial', but this is a rather curious controversy. The evidence that there is some inheritance of epigenetic marks seems increasingly incontrovertible (see e.g. Morgan et al. 1999: Tang et al. 2015), and most of the argument seems now to be about the robustness of such inheritance over multiple generations.

It is hard to avoid the suspicion that much of this controversy is ideological. Supporters of epigenetic inheritance occasionally suggest that this will provide an opportunity for 'Lamarckian' inheritance (Jablonka and Lamb 1995), the ultimate prohibition in orthodox Darwinian thought. And while this is certainly possible, there is no obvious connection between epigenetic inheritance and Lamarckianism.[8] A more

---

[8] Understood here in the somewhat vulgar sense as any inheritance of acquired characteristics that are acquired because they enhance fitness. Epigenetic changes are indeed acquired, but whether they are acquired because they are adaptive is another matter. If the reason that an organism acquires adaptive

important point is that the inheritance of epigenetic characteristics need in no way depend on transmission through the germline. The best illustration of this is the paradigm for behavioural epigenetics, the work of Michael Meaney and collaborators on maternal care in rats (Champagne et al. 2006). Meaney showed that proper care affected the developing brains of rat pups, making them less liable to be fearful and also, crucially for female pups, more likely to provide proper care to their own offspring. Although there is some controversy surrounding Meaney's interpretation of his results, at the very least the example illustrates how it is entirely possible in principle for an epigenetic trait to be inherited via the epigenetic effect of its behavioural manifestation.

If this seems unlikely to be a widespread phenomenon, it is at least worth reflecting on the ubiquity of epigenetic expression in the brain. Any experience that affects an organism in any lasting way will, presumably, leave some physical change in their brain, and this, we generally suppose, will involve the expression of genes and the production of proteins. Triggering such a process is what epigenetic modifications to, or chemical alterations of, the genome do. In a reasonably broad sense of 'epigenetics', the brain is undergoing epigenetic modification moment by moment.[9] That these modifications should be involved in more or less elaborate processes that enable their passage to other organisms, including their offspring, should not be surprising. In fact, of course, we know well that it happens all the time, and most spectacularly in our own species, by means of socially mediated learning.

## 5.5 Learning

Evolutionary psychologists tend to become very exercised when it is suggested that learning presents a problem for their views. They suppose that learning is taken to be some universal, content-independent way of acquiring any arbitrarily selected bit of competence or information, whereas in fact it is the exercise of a very specifically directed capacity. Thus Cosmides and Tooby (unpublished) write: 'Students often ask whether a behavior was caused by "instinct" or "learning". A better question would be "which instincts caused the learning?"' Later they continue:

> The cognitive architecture, like all aspects of the phenotype from molars to memory circuits, is the joint product of genes and environment. But the development of architecture is buffered against both genetic and environmental insults, such that it reliably develops across the (ancestrally) normal range of human environments. [Evolutionary Psychologists] do not assume that genes play a more important role in development than the environment does, or that 'innate factors' are more important than 'learning'. Instead, [Evolutionary Psychologists] reject these dichotomies as ill-conceived.

---

characteristics is that it has evolved a capacity to do so by orthodox Darwinian means, this would not be a significant threat to the orthodoxy.

[9] It is suggestive that, according to Tang et al. (2015), the parts of the genome that resist demethylation of germ cells are disproportionately expressed in the brain.

Indeed, in terms of learning, we cannot even sensibly ask which of genes or the environment is the more important. For:

Any developmental biologist knows that this is a meaningless question. Every aspect of an organism's phenotype is the joint product of its genes and its environment. To ask which is more important is like asking, Which is more important in determining the area of a rectangle, the length or the width?

Learning, then, is a way in which the environment contributes to the production of the universal human cognitive architecture. Genes and environment are inextricably linked in this process, and the reason it makes no sense to separate their contributions is because they are both doing exactly the same thing: jointly producing the universal species-typical outcome. Now certainly I don't want to deny that there is a good deal of buffering in development. Without this, it is hard to imagine any successful development happening at all. But surely learning would be a very strange device to evolve merely to buffer development against environmental noise. In fact, Cosmides and Tooby do allow that learning may function as a kind of switch, determining which of two developmental pathways, both selected by past evolution, an organism will follow. No doubt they would allow a predetermined range of developmental trajectories. But even this takes us only to the level of plasticity I described with examples from social insects; it does not begin to capture the functional plasticity found in higher mammals and, especially, humans.

Here is where the universality of plasticity should lead us to suspect that we are going down a wrong path. Certainly development must produce a sufficiently constant product to reliably meet the conditions of existence. This is no mean feat. But successful lineages, it appears, manage to embed within this reliable reproduction of their kind a great deal of flexibility that enables at least fine-tuning to the particular exigencies of the ever-changing and always differentiated environment, and to do so on a timescale far more rapid than could result from the natural selection of randomly generated genetic mutations. Whether, in the spirit of the neutral theory, we should see the latter as mainly irrelevant to the processes of adaptation is a matter for another time. But what is important now is that organisms have a wide variety of resources—developmental, metabolic, behavioural, and no doubt others—that have evolved precisely to facilitate adaptation at more or less rapid timescales. The human mind is a spectacular example of such an evolved set of capacities.

## 5.6 Niche Construction

More or less intelligent behaviour is of obvious importance in sustaining the lives of individual humans, but it plays an equally important part in the stabilization of human lineages.[10] Most obviously, the rearing and educating of children is a domain

---

[10] I shall not address here the interesting question of the relevant scope of a human lineage. My assumption is that there is no unique answer: humans are socially integrated at multiple scales, with variable

of behaviour essential to the perpetuation of any human lineage. The processes by which behaviour is transmitted between generations has been extensively investigated by students of cultural evolution (see e.g. Richerson and Boyd 2005). But perhaps even more intriguing than these vital processes is the production and maintenance of so-called developmental scaffolding (Bickhard 1992; Sterelny 2003) or developmental niche construction. The latter, according to Karola Stotz (2010), asks

not 'what is inside the genes you inherited', but 'what the inherited genes are inside of' and with which they form a wider whole—their internal and external ontogenetic niche, understood as the set of epigenetic, social, ecological, epistemic and symbolic legacies inherited by the organism as necessary developmental resources.

Developmental niche construction began with an idea that goes back at least to Darwin's (1881) treatise on earthworms, and that has been brought back into prominence by John Odling-Smee, Kevin Laland, and Marcus Feldman (2003). Organisms do not just respond to the environment that they find, but devote a great deal of energy to transforming the environment, generally in ways to which they are adapted or in ways that promote the development of their offspring (or other younger members of their species).

Of particular importance to the human case is scaffolding or niche construction that promotes cognitive development, and indeed the lives of juvenile humans are largely structured around their cognitive development. Schools and a host of associated learning technologies are the most obvious structures developed to this end, but many informal interactions between children and parents, other adults, and peers are more or less explicitly focused on cognitive development. The physical environment in which humans develop and live, moreover, is structured in many subtle ways to facilitate intelligent activity. Think of an everyday site of complex activity, the kitchen. Heights of surfaces, spatial relations between appliances and work surfaces, the sizes and dispensing devices of containers for ingredients, the design and positioning of utensils, and so on are all more or less well designed to guide and facilitate effective action in the execution of complex food-preparation tasks.

There is of course much more that could be said on this topic. My present point, however, is a relatively simple one. This active shaping of the environment to promote cognitive and hence behavioural effectiveness must be understood in terms of the organism as open-ended, developing in a partially unpredictable way, rather than as a thing with a fixed repertoire of species-universal dispositions.[11] Perhaps the best way of seeing this is to see how this claim locates the sound middle ground between the extremes of a blank slate and a mind programmed to produce predetermined responses to given situations. In terms of the latter, if that were the kind of cognitive development that were required, it could be innate, in the sense of appearing largely irrespective of

---

degrees of permeability between these integrated groups. Proper attention to this complication would, no doubt, add a significant layer of complexity to the discussion.

[11] This remark of course reflects my broader theme: that the human organism should be seen as a process rather than as a substance.

contingencies of the environment, and this would save the adult phases of the organism an enormous amount of effort. But on the other hand, as evolutionary psychologists like to insist, a blank slate would require far too much learning; trial and error would far too often lead the developing organism into fatal error. The solution is obvious: the 'blank slate' is constrained by the structures that are provided by parents and other conspecifics, living or dead. Human children, to take a trivial example, do not learn to cross roads by randomly trying a variety of techniques. The cost of failure would be too high. They are taught to do this by older humans. And this is just as well, since there were no roads in the so-called environment of evolutionary adaptation, and if they relied on past evolution they would mostly be dead. The important point about this strategy is that it enables evolution of the niche, in this case the invention of roads. This requires both cognitive creativity—someone has to think of roads—and cognitive adaptation—our children have to learn to survive in a road-rich environment. Neither a blank slate nor a universal human cognitive architecture will do these jobs.

## 5.7 Back to Human Nature

There is much to be said for a flat-out denial that there is any such thing as human nature. The concept suggests something fixed, whereas, as I have argued, human behaviour is an adaptation for permitting plastic responses to a changing environment—an environment that may even be changed by the adaptive activities of humans themselves.

There is nothing in principle objectionable about a concept of human nature that consists of no more than the description of what humans at a particular time and place typically do—a kind of localized natural history. This is presumably the kind of thing that anthropologists, sociologists, or historians often try to provide, and I certainly don't mean to object to their practices. Such social scientists, while not generally committed to what evolutionary psychologists refer to as the 'standard social science model'—the blank slate—are certainly aware that the local natural histories they provide would not apply precisely, or sometimes even closely, to humans at distant times and places. On the whole, it seems best not to risk confusion of such local descriptive ventures with the much more portentous traditional concept of human nature.

I also advocate a stronger reason for rejecting the concept of human nature: that I take humans to be better understood as processes than as things or substances. The nature of a thing serves, among other functions, to determine what it is for a thing to persist. It does so for as long as it continues to realize the nature of the kind of thing it is. But if only because organisms undergo developmental pathways that may lead them through radically different states, this cannot be the kind of nature they have. It is true that human nature is generally understood as referring rather specifically to the behavioural or the cognitive, and often to these in relation to the adult. But here there is a clear tendency for such language to mislead. Attributing the nature of an organism to one particular developmental stage inevitably distracts attention from the process by which it reached that developmental stage; and it is not news that popular theories of human nature are often derived from evolutionary theories rightly criticized for their neglect of development. If development is, as is surely the case, a substantially plastic process, then overlooking it and attributing an invariant state to the developmental outcome is clearly misguided. Seeing the organism itself as a process, the intersection of many constituent processes that together reliably produce a stable structure that satisfies the conditions of its existence, makes errors of this kind impossible.

C. H. Waddington, perhaps the most important of the mid-twentieth-century biologists strongly influenced by Whitehead, distinguished between homeostatic and homeorhetic processes. Whereas the former referred to processes that maintained a constant condition of a system, the latter referred to the processes constraining a system to a particular pathway. Clearly both kinds of processes are exhibited by organisms, development being a central example of a homeorhetic process.[12] Could one then perhaps say that humans were distinguished by a particular species-specific set of homeorhetic processes, and the typically reproduced stages thus generated constituted the nature of the human (process)? If to characterize the nature of a kind of entity is just to describe it as well as possible, then indeed this is what one could and should say. However, I suggest that there is a great deal of baggage connected to the concept of human nature that is strongly at odds with the kind of account just indicated. It is not merely that a developmental process need not exhibit any characteristics that are constant throughout its career, but also that the plasticity I have stressed implies that—especially with respect to the psychological traits most generally associated with human nature—there is a lot of variability in the features manifested at particular stages of the process. All this is so far from the common associations with the expression 'human nature' that I think it is safer to dispense with its use altogether.

I have referred at several points to processes that maintain a stable structure. But note that I am not setting up the structure as something fixed and opposed to the functional or processual. As J. S. Haldane (1931: 22) summarized this crucial feature of a processual understanding of biology: 'Structure and functional relation to environment cannot be separated in the serious scientific study of life, since structure expresses the maintenance of function, and function expresses the maintenance of structure.' This thought beautifully captures the decisive shift from the organism as a thing with a certain set of properties conducive to its adaptation, to a process constantly acting to maintain its integrity and persistence. And from the latter viewpoint, adaptive plasticity, or flexible adaptation, is what behaviour is always aspiring to achieve.

---

[12] Note, however, that this classification of processes is, in an important way, relative to perspective and temporal scale. From an evolutionary perspective, development might better be described as a homeostatic process in which cycles of development spin off new cycles through reproduction, thereby stabilizing the lineage. Similarly, metabolic cycles that homeostatically maintain features of the cell are themselves, in finer detail, homeorhetic sequences of chemical reactions.

# References

Bickhard, M. H. (1992). 'Scaffolding and Self-Scaffolding: Central Aspects of Development.' In L. T. Winegar and J. Valsiner (eds), *Children's Development within Social Contexts: Research and Methodology*, 33–52. Mahwah, NJ: Erlbaum.

Champagne, F. A., Weaver, I. C., Diorio, J., Dymov, S., Szyf, M., and Meaney, M. J. (2006). 'Maternal Care Associated with Methylation of the Estrogen Receptor-Alpha1b Promoter and Estrogen Receptor-Alpha Expression in the Medial Preoptic Area of Female Offspring.' *Endocrinology* 147: 2909–15.

Cosmides, L., and Tooby, J. (unpublished). 'Evolutionary Psychology: A Primer': www.cep.ucsb.edu/primer.html

Darwin, C. (1881). *The Formation of Vegetable Mould through the Action of Worms, with Observations on their Habits*. London: John Murray.

Dupré, J. (2001). *Human Nature and the Limits of Science*. Oxford: Oxford University Press.

Dupré, J. (2008). 'Against Maladaptationism: or What's Wrong with Evolutionary Psychology?' In M. Mazzotti (ed.), *Knowledge as Social Order: Rethinking the Sociology of Barry Barnes*, 165–80. Farnham: Ashgate.

Dupré, J. (2010). 'The Conditions for Existence: Review of John O. Reiss, Not by Design: Retiring Darwin's Watchmaker.' *American Scientist* 98: 170–72.

Glennan, S. S. (1996). 'Mechanisms and the Nature of Causation.' *Erkenntnis* 44: 49–71.

Haldane, J. S. (1931). *The Philosophical Basis of Biology*. London: Hodder & Stoughton.

Jablonka, E., and Lamb, M. J. (1995). *Epigenetic Inheritance and Evolution: The Lamarckian Dimension*. Oxford: Oxford University Press.

Keijzer, F. (2013). 'The Sphex Story: How the Cognitive Sciences Kept Repeating an Old and Questionable Anecdote.' *Philosophical Psychology* 26: 502–19.

Lewens, T. (2015). *The Meaning of Science*. London: Pelican.

Machamer, P., Darden, L., and Craver, C. F. (2000). 'Thinking about Mechanisms.' *Philosophy of Science* 67: 1–25.

Machery, E. (2008). 'A Plea for Human Nature.' *Philosophical Psychology* 21: 321–9.

Morgan, H. D., Sutherland, H. G. E., Martin, D. I. K., and Whitelaw, E. (1999). 'Epigenetic Inheritance at the Agouti Locus in the Mouse.' *Nature Genetics* 23: 314–18.

Nicholson, D. J. (2012). 'The Concept of Mechanism in Biology.' *Studies in History and Philosophy of Biological and Biomedical Sciences* 43: 152–63.

Nicholson, D. J. (2013). 'Organisms ≠ Machines.' *Studies in History and Philosophy of Biological and Biomedical Sciences* 44: 669–78.

Odling-Smee, F. J., Laland, K. N., and Feldman, M. W. (2003). *Niche Construction: The Neglected Process in Evolution*. Princeton, NJ: Princeton University Press.

Reiss, J. (2009). *Not by Design: Retiring Darwin's Watchmaker*. Berkeley: University of California Press.

Rescher, N. (1996). *Process Metaphysics: An Introduction to Process Philosophy*. New York: SUNY Press.

Richerson, P. J., and Boyd, R. (2005). *Not by Genes Alone: How Culture Transformed Human Evolution*. Chicago: Chicago University Press.

Sterelny, K. (2003). *Thought in a Hostile World*. Oxford: Blackwell.

Stotz, K. (2010). 'Human Nature and Cognitive-Developmental Niche Construction.' *Phenomenology and the Cognitive Sciences* 9: 483–501.

Tang, W. W. C., Dietmann, S., Irie, N., Leitch, H. G., Floros, V. I., Bradshaw, C. R., Hackett, J. A., Chinnery, P. F., and Surani, M. A. (2015). 'A Unique Gene Regulatory Network Resets the Human Germline Epigenome for Development.' *Cell* 161: 1453–67.

West-Eberhard, M.-J. (2003). *Developmental Plasticity and Evolution*. New York: Oxford University Press.

Whitehead, A. N. (1978). *Process and Reality: An Essay in Cosmology*. New York: Macmillan.

# 6

# Sceptical Reflections on Human Nature

*Kim Sterelny*

## 6.1 Hull's Challenge

In his 'On Human Nature', David Hull (1986) argued that the very idea of human nature was tied to an essentialist concept of biological species, one that failed to recognize both the variability and the dynamism of biological kinds.* Beginning with Darwin, and with even greater emphasis in more recent evolutionary theory, evolutionary biology has taught us that species are typically variable at a time, that they change over time, and that there is no principled bound on the variation to be found in a species at a time, or over time.[1] Hull took this point to be reinforced by his view that species are individuals rather than kinds. The parts that make up individuals need not be similar to one another in any distinctive way. My left thumb and my liver are both parts of me, but not in virtue of any similarities they may have. But while the idea that species are built through distinctive historical and genealogical processes remains firmly part of biological consensus (recently and elegantly reformulated in Ereshefsky 2014), that is much less true of the idea that species are individuals. As, for example, Samir Okasha (2002) and Peter Godfrey-Smith (2014) have noted, the biological insights of the idea that species are built by a distinctive evolutionary process in no way depends on the metaphysical claim that they are individuals. Even so, the point that there are no in-principle bounds on the extent of variation remains important, and our species

---

* Thanks to Tim Lewens and Steve Downes for their feedback on an earlier version of this chapter, and to the audience at Victoria University of Wellington, likewise, for their feedback. Thanks also to the Australian Research Council for a series of generous grants that support my research on issues connected to human social and cognitive evolution.

[1] There was a time when one might have been tempted to read Gould and Eldredge's (1993) 'punctuated equilibrium' model of species life history as a kind of essentialist counter-revolution, seeing species phenotypes as actively stabilized over the lifetime of a species, with a species phenotype depending on an integrated genetic system that could breakdown only under special circumstances. In their final joint paper on these ideas, Gould and Eldredge made it clear that this was not, and never had been, their message. Species phenotypes are stable because change does not typically accumulate, not because they are stabilized by intrinsic genetic or developmental mechanisms.

is no exception to this message about the importance of variation. While there is relatively little genetic diversity in our species (since *H. sapiens* is a young species), individual plasticity is very marked; in particular, our cognitive, behavioural, and social phenotypes are strongly responsive to both individual and social learning. Human variation is so obvious and so salient that, much to the displeasure of nativist evolutionary psychology, the main focus of the social sciences has been to explain change over time and variation at a time.

As a consequence of these considerations, Hull argued that there was no interesting and important characteristic of humans (nor even any cluster of characteristics) that was both genuinely and strictly universal, and distinctive of our species; recall that, to be distinctive, a nature-defining characteristic must characterize our species but not the recently extinct members of our clade. Moreover, as Hull goes on to argue, even if there were some trait that was both universal and distinctive, there is no guarantee that it would be universal and distinctive amongst our changed but still conspecific descendants. It would cease to be distinctive if our lineage split; it would cease to be universal if our evolving lineage developed variation in that trait, and, as emphasized above, there is no a priori limit to the amount or kind of variation that can emerge in an evolving lineage.

One response to Hull is to meet his challenge head on, rejecting his conception of species and essences. Michael Devitt (2008, 2010) has attempted an essentialist counter-revolution, arguing that species do have (partly) intrinsic essences. In his view, this follows from the fact that species have characteristic, if not invariable, essences. Being told Sakti is a tiger is not just informative; it is explanatory: Sakti's being a tiger explains Sakti's stripes: 'the fact that an individual organism is a tiger, an Indian rhino, an ivy plant [...] explains a whole lot about its morphology, physiology and behavior' (Devitt 2008: 352). Now even if Devitt were right, on this revived version of essentialism, our more distinctive and impressive phenotypic characteristics would not be essential to being human. The essential properties are the genetic and developmental bases of these phenotypic traits. It is always possible for environmental misfortune to block normal development. So human nature could not be identified with the biological essence of being human; perhaps, though, it could be identified with its normal phenotypic expression. Understanding and responding to moral norms could not be an essential feature of being human in this sense, and hence not part of an essential nature, even if moral psychology were to have a distinctive genetic basis unique to our species. For there would be individuals with the genetic basis of moral psychology, but who developed no sensitivity to norms; perhaps they grew up in very unusual environments or suffered developmental misfire.

However, Devitt's counter-revolution, rightly, has not changed the anti-essentialist consensus in philosophy of biology.[2] In my view, the fact that Sakti is a member of

---

[2] More exactly, an essentialism that appeals to similarity based on intrinsic properties. One response to a historical conception of species and speciation is to claim that the essence of being a tiger is the relational

*Panthera tigris* does *not* explain why Sakti has stripes.³ That explanation is developmental, not genealogical. To explain Sakti's stripes, we need to identify the developmental mechanisms that control tiger fur-patterning. Up to a point, Devitt agrees. He is happy to distinguish between historical and structural explanations.⁴ But he insists that the structural explanation must appeal to an essential property—a tiger-defining property—of the developing animal. He says, 'at bottom, structural explanations will advert to essential, intrinsic, largely genetic, properties' (Devitt 2008: 354), and sums up his whole thesis as the claim that 'structural explanations in biology demand that kinds have essential intrinsic properties' (p. 355).

While not uncontroversial,⁵ the claim that developmental explanations depend on intrinsic, largely genetic properties is plausible. But there is no reason to suppose that these intrinsic properties are essential, and good reason to doubt it. For the power of the structural explanation in no way depends on the distribution of the developmental resources to which it appeals. Consider a developmental biologist working on the development of the tooth morphology of the tiger canine. The biologist herself, or another developmentalist reviewing her paper and the explanation it contains, would not be able to tell, even from an ideally complete account of the mechanism, whether the tooth growth pattern was distinctive to tigers, to the larger felid clade in which tigers are embedded, or to Sumatran tigers, the subspecies to which Sakti belongs. The structural explanation of Sakti's tooth morphology is good or bad, independently of whether the same explanation applies to lions and leopards, or whether this is a tiger apomorphism. Likewise, the power of the explanation is unaffected by the possibility that there are alternative developmental pathways to the same morphology. Within the sea urchin clade, there are alternative developmental pathways to the radial symmetry of adult urchins (Raff 1996: 223-37). While to the best of my knowledge these do not vary within the population of living species, presumably there was variation within single populations through evolutionary transitions between direct and indirect development of radial symmetry. A historical explanation specifies the distribution of developmental resources through the felids; a structural explanation tells us, given that an animal has a particular package of developmental resources, how they build the animal's phenotype. The character of that structural explanation in no way depends on whether that package is sometimes, but not uniformly, present in the population; on whether it is characteristic of this population, but not of all populations; or on whether it is characteristic of the whole species or of some larger taxon.

---

property of being genealogically integrated within the tiger lineage, but Marc Ereshefsky (2010) suggests that while this version of essentialism is not clearly mistaken by the lights of evolutionary theory, nor is it mandatory.

³ As Tim Lewens (2012a, 2012b) shows, it is possible to accept Devitt's view that species membership explains an organism's manifestation of its characteristic phenotype, while resisting his essentialist conclusion.

⁴ This is his terminology for Mayr's proximate/ultimate distinction.

⁵ It would be rejected by developmental systems theorists and their allies; see e.g. Oyama et al. (2001).

Devitt accepts that the idea that the intrinsic properties must be essential requires defence (Devitt 2008: n. 21). But when he comes to defend it at the very end of the paper, he just claims that the causal chain from the developmental basis of stripes to Sakti having stripes would not be lawlike unless that developmental basis were an essential property of tigers. Now perhaps it is true that the causal chain from Sakti being a tiger to Sakti having stripes would not be lawlike unless the intrinsic basis of Sakti's stripes were an essential property of tigers. But, of course, anti-essentialists do not think it is a law, or anything like a law, that tigers are striped. Colour patterns are notoriously developmentally and evolutionarily labile. Moreover,[6] the mechanistic, structural explanation of the development of striping can be thoroughly lawlike. It is just that the laws will be about molecular mechanisms, cell differentiation, biochemical gradients, protein structures, and the like. Tigers will not get a mention.

Finally, I doubt that Devitt has answered the standard objections to intrinsic essentialism: that there is no principled upper bound to the extent of variation in a species at a time or over time. In principle, there could be very extensive variation within the same species, so long as the potential remains for the genes in any two members to be combined in a downstream generation. As far as I can see, Devitt just denies this outright. While he accepts that there is, of course, no 'tiger gene', he just insists that there is 'a certain cluster or pattern of underlying, largely genetic, properties [...] common and peculiar to tigers' (2008: 371). It is probably true that in most real cases, the extent of variation at a time is limited; this is why systematicists often regard a threshold percentage sequence difference between sister lineages as a speciation signal. Devitt's essentialism does not commit him to the literal existence of a tiger gene. Whether something like a cluster model of genetic similarity can bear the theoretical weight Devitt's analysis demands is an open question (Lewens 2012a). More decisively, lineage-splitting is bad news for the idea that a species essence is a cluster of genetic characteristics. Take a candidate, tiger-defining essence, say, the set of genetic resources $F$. Imagine the tiger lineage-splitting; the Sumatran subspecies evolves a lion-like social life, and as a consequence becomes reproductively isolated from the main *tigris* lineage, as females in the Sumatran lineage breed only with males in coalitions. These Sumatrans are very likely to still have $F$ as part of their genome (perhaps some elements within $F$ are genetically masked—a common form of evolutionary change—or perhaps the newly evolved social behaviour depends on genetic changes outside $F$). The underlying genetic properties would still be common, but no longer peculiar to tigers.

## 6.2 Human Nature Without Essentialism

In short, the counter-revolution fails and Hull's anti-essentialism is still received wisdom in thinking about species. Even so, Hull's own arguments emphasized the lack of

---

[6] To the extent that we think mechanistic explanations are lawlike. I myself see mechanistic explanation as an alternative to that conception of scientific explanation.

plausibly important and strictly universal phenotypic traits, and there is an obvious response to this line of thought. Consider the plausible candidates for being part of human nature: perhaps the ability to speak and understand language, or the ability to consciously and deliberately reason about the world beyond the limits of perception. It is true that some biological members of our species fail to develop these capacities. Some die too young; others suffer from crippling pathologies or profoundly impoverished developmental environments. But all humans have *the potential* to develop such capacities, even the ones who died young, or suffered genetic or developmental misfortune. So all normal, or normally developing, members of our species do have these traits.

Hull was quite unpersuaded by this response, dismissing them as question-begging evasions. He denied that there was a robust, untainted notion of 'normal development', and denied that there was a well-defined sense in which a human suffering from a language-disabling mutation nonetheless had the potential to develop language. By contemporary lights, Hull overstates his case. Consider language and a genetic or developmental anomaly that blocks the development of language in some member of our species. That individual still has the potential for language in a sense in which, say, a chimp lacks that potential. In the case of a mutation-driven failure, had that human had any of a range of common human alleles at the site of the mutation or mutations, he or she would have developed the capacity to use language. A chimp would not, even if the chimp genome has sites homologous to the mutated sites: chimps do not have the requisite background genetic and developmental resources, equally essential to the development of language. Imagine, for example, that the FOXP2 gene really had the language-specific effects that its discovery initially suggested. Even had that been the case, had a human-typical form been inserted into a chimp,[7] and had that chimp then been raised in a human family, it would not have developed mastery of that human's language. Perform the same manipulation on an otherwise genetically typical human child with a mutated FOXP2, and the human child in that same environment would master language. Many factors have to be in place (we can be sure) for language to develop, and much of that is missing in the chimp. So a human lacking one (or a few) of the many necessary ingredients for language is a potential speaker, in the way a chimp is not. Obviously, as the number of defective ingredients goes up, the less there is any such potential. So there will be indeterminate cases, but perhaps that is no problem.

Moreover, there are independently plausible accounts of function and malfunction that do not smuggle in political, social, or moral norms (Godfrey-Smith 1993, 1994). It is very likely that these will count genetic and developmental histories that derail, say, the development of language as failures of normal development. Normal humans acquire language. Furthermore, as Edouard Machery (2008) has pointed out, there seems no reason why a theory of human nature should commit itself to the claim that critical, core human characteristics are *strictly universal*. Perhaps the characteristics that apparently make us so unlike other great apes—our sociability, our distinctive technical

---

[7] Chimps have a somewhat different version of the gene; see Dominguez and Rakic (2009).

capacities, language, religious and/or moral norms, our awareness of one another as thinking agents—are distinctive and typical. So understood, as Machery points out, an account of human nature will not specify the necessary and sufficient conditions for being a member of our species. But genealogical connection does that. Godfrey-Smith has made a similar point: 'once evolution has taken a lineage in a particular path [...] we can speak about an "evolved nature" [...] in that lineage' (Godfrey-Smith 2014: 142), even though the particular traits may not be strictly universal, and even though some may eventually change. As he points out, variation within species does not make the production of field guides impossible. That would probably be true even if field guides relied only on developmentally canalized traits.

So Hull frames claims about human nature narrowly, as a claim about the biological character of the species of which living humans are members, *Homo sapiens*; and even given that narrow frame of reference, his arguments as stated are over-drawn. Machery and Godfrey-Smith show that we can give an account of human nature consistent with the anti-essentialist consensus.[8] But his challenge is deeper than his arguments. Once we recognize the historical and genealogical nature of species in general and our species in particular (so to be a human is just to be reproductively embedded in the right lineage), once we recognize that change over time and variation at a time matter (so that variation and change are not noise to be idealized away), it is not clear we need a theory of human nature.

Suppose, for example, that typical humans (a) remember themselves in various past circumstances, fortunate and unfortunate, and can imagine themselves in different future scenarios; (b) can use language in a flexible and open-ended way; (c) can reason about unobserved forces, events, and agents; (d) understand that others are also thinking and feeling agents; (e) make normative judgements about their own acts and those of others; and (f) understand and are anxious about their own mortality. Suppose it is also true that no other species has this combination of characteristics, and that this combination is typical not just of contemporary humans, but of humans through the existence of our species.

This is a coherent account, though not one that would serve for a field guide, since these traits are not readily detected by chance observation. How can we judge whether this would count as an adequate theory of human nature, and if it did not, in what way would it be a failure? After all, I have presented above is no more than a list, and we need some principled account for inclusion in that list. The ability to laugh; the capacity to feel moralized emotions like guilt, shame, or pride; a delight in telling and listening to stories; the use of fire are all as prima facie plausible candidates for inclusion as those specified above. Should they be there too? If so, it will not take long to think up some more plausible candidates—traits that emerged deep enough in our history, and

---

[8] Lewens (2012a, 2012b) is not convinced. These accounts require the elements of our evolved nature to be both typical and evolved, and Lewens suggests that in the light of our developmental plasticity and sensitive to cultural input, it is by no means clear that evolved traits can be distinguished from others.

were persistent enough, once they emerged, to be widely distributed across human cultures and typical in those cultures. Even if we think of this project as simply descriptive, aiming at no more than a characterization of the typical, generic human, there would still be serious issues about what to include and how to characterize those traits. For example, anthropologists debate whether there is a generic specification of our emotional repertoire, or whether emotions are so entwined with the specific features of a given culture that they cannot be cross-culturally labelled. Moreover, many of these traits are older than our species—the use of fire certainly is—so such a list is more a specification of the typical traits of recent species in our clade than a specification of *sapiens*, specifically. Perhaps it is the cluster itself—having all of these traits—that is specific to our species. That said, discussions of human nature are often somewhat ambiguous between a narrow reading, targeting our specific species, and a broader characterization of our lineage.

To answer this challenge, we need to know why, if at all, we need a theory of human nature. A portrait of the typical human can be built in many, very different ways. Without a conception of the theoretical work a theory of human nature should do, we have no way of determining the kinds of traits we should specify (adult traits? children's traits? life-history pathways like menopause or life expectancy?), and no way of determining just how typical of us a trait should be. Given the population growth of the last centuries, the typical human has probably lived in the last century or so. But should 'recently dead' be part of the portrait of statistical typicality? We need to know what a concept of human nature is for, and there seem to be three natural suggestions. (i) A theory of human nature specifies an explanatory project; it characterizes the unique features of our lineage, and hence the target of an evolutionary account of human uniqueness. (ii) Humans have lived in an impressively diverse set of social arrangements. But presumably that actual and potential array is both constrained and biased in various ways. For example, it is very plausible indeed that social peace and cooperation have to be managed differently in large- and small-scale social worlds. Perhaps a theory of human nature enables us to characterize the space of possible human social worlds, and of possible pathways from one social world to another. (iii) Our normative attitudes to other humans are very different from our attitude to non-human animals. Most obviously, other humans have legal and moral rights that non-humans lack. If those differences in normative attitude are to be rationally defended, humans and non-humans must be different in some important and relevant way. Perhaps a theory of human nature is a theory of that normatively relevant difference.

## 6.3 Human Nature as the X-Factor

I have argued before that the human evolutionary career—in the sense of the hominin clade, rather than our specific species—does pose a special problem for evolutionary biology. We are very different from our closest great ape relatives in our social lives, our bipedal gait, our reproductive organizsation, our technical skills, our communicative capacities, and our forms of coordination and cooperation (Sterelny 2007, 2012). These differences have evolved very rapidly. We shared a common ancestor with the chimp clade only six or seven million years ago; our ecological and social take-off was probably more recent still. So there is a genuine question about human uniqueness: what explains our extreme and rapid divergence from the great ape stocks?

There has been a popular strategy in responding to this question: the 'key adaptation' model. The idea is to identify a critical, game-changing adaptation in our lineage, one that transformed our ancestors and their way of life, thus leading to a cascade of other changes. Unsurprisingly, given the list in section 6.1, there are plenty of candidates for this critical adaptation. These include: language, cognitive flexibility, and the capacity to integrate information across distinct domains; enhanced cultural learning and the ability to accumulate cultural information across generations; enhanced working memory and the capacity to plan; and emotional and motivational changes that made us more tolerant and cooperative.[9]

If any hypothesis along these lines is correct, then the list-like characterization of human nature is unsatisfactory, for most of the items on the list will be downstream consequences of the crucial adaptive breakthrough. The list does not make salient the fact that one change is fundamental and responsible for the others. For example, if Stephen Mithen (1996, 2013) is right, and cognitive integration led to a revolution in human thinking, social life, and ecological footprint, then integration is 'the difference that made the difference'. Humans are distinctive in having integrated, rather than rigidly modular, minds. These evolutionary models of human uniqueness make sense of the search for human nature, seen as a distinctive, difference-making feature of our clade; what makes us special is that difference-maker.

However, I am deeply sceptical of this whole strategy. I am one of those who think that the 'key innovation' model of hominin evolution is quite mistaken,[10] and I have developed an alternative framework in which the hominin evolutionary trajectory is explained by appeal to positive feedback between reproductive cooperation, information sharing, technical expertise, and the collaborative exploitation of valuable but difficult-to-exploit resources, coupled with an initial trigger that pushed our lineage into a distinctive evolutionary trajectory. On this view, there is no key innovation or threshold-crossing event when our quasi-human ancestors became fully human. Instead, there is a mix of factors. Some are innate individual characteristics: our bipedal gait; the metabolic costs, and birthing risks, of human babies; the extreme dependence and great plasticity of human infants. Some are individual traits, but ones that probably develop very differently in different sociocultural settings: technical expertise, normative thought. Some are features of the social environment rather than of individual

---

[9] For a sample of these key innovation models, see Mithen (1996), Tomasello (1999), Klein and Edgaar (2002), Hrdy (2009), and Wrangham (2009).

[10] Sam Bowles, Herb Gintis, Robert Boyd, Peter Richerson, Joe Henrich, and (more recently) Michael Tomasello also propose coevolutionary models of various kinds; see e.g. Henrich (2004, 2006), Bowles and Gintis (2011), Boyd et al. (2011), Tomasello (2014), and Boyd and Richerson (2013).

agents within that environment. Cooperation between parents and between a mother and her kin in supporting and educating her children is not an individual trait but is a feature of a human social world, and one that is, arguably, both historically deep (Hawkes et al. 1998; Hrdy 2009) and very important in increasing the robustness and reliability of social learning (Burkart et al. 2009).

One could just say that 'human nature' simply names this whole matrix of features: the combination whose coevolution explains the distinctive features of our life. Human nature is just the package of typical human traits. But that would obscure rather than clarify. For it masks the difference between a coevolutionary and a key innovation model. To succeed, a coevolutionary model must (i) specify an early hominin baseline: a set of cognitive, social, and physical capacities that early hominins had, shortly after their separation from the great ape lineage; (ii) identify the initial coevolutionary interactions that began to drive divergence from that baseline; (iii) identify a sequence of incremental changes in hominin morphology, social organization, behaviour, cognition, and ecological role, beginning from those baseline capacities and linking to those manifest in our species; (iv) give a plausible account of the selective advantage of each of those changes, in concert with the others. In any account with this shape, the specific form of a trait, and its relative importance, is likely to change through this trajectory, and traits that are now of fundamental importance to our social lives (normative thought, language, division of labour) may well have appeared quite late. Conversely, others that are not now central to the social lives of most humans might once have been pivotal to our existence. On some views, shared song and coordinated dance were once pivotal in making group life possible (Mithen 2009; Morley 2013). That role is now largely supplanted by formal institutions. If a narrative of this form explains the distinctive character of our lives and minds, it would be misleading to say that our evolved human nature explains why we are so different from great apes. This would not be so if (say) self-control and planning evolved for local and contingent reasons in some hominin population, and that resulted in the hominin revolution (Wynn and Coolidge 2012). For then conscious planning—executive control—would be the human X-factor. These coevolutionary accounts also erode a robust distinction between biologically caused and culturally caused features of typical human lives. This distinction, of course, was already problematic in human developmental sciences, as the innate/acquired distinction is now rightly seen as problematic (Mameli and Bateson 2011). These coevolutionary views make the distinction problematic for historical, ultimate explanations as well. Our long history as animals dependent on social learning undermines a distinction between biological (genetic) and cultural-historical explanations.

One might, of course, take human nature to be the target of these evolutionary explanations, rather than as an explanatory resource. On that proposal, 'human nature' is just a name for the very distinctive features of our lineage that require special explanation. While there is nothing clearly wrong with that suggestion, it is not the most helpful way of framing the explanatory agenda. For many of the most puzzling features of human evolution are not individual traits but aspects of our social life and population structure. One example is the fact that we are by far the most numerous and widely distributed large animal, and have been since the late Pleistocene. Is cosmopolitan distribution an aspect of human nature?

## 6.4 Possible and Plausible Worlds

Somewhat notoriously, appeals to our evolved nature have played a role in debates about feasibility—about humanly possible social worlds. The idea depends on something like Machery's 'nomological conception' of human nature: our typical physical, cognitive, and motivational traits constrain human social possibilities. In a few famous cases, this line of thought has been yoked to conservative moral and political agendas.[11] It thus helped make early sociobiology and many forms of evolutionary psychology an anathema to those on the left (for vigorous scepticism about this appeal to human nature, see Kitcher 1985; Dupré 2001). How bad are these ideas? No doubt Wilsonian sociobiology was probably too sympathetic to muted forms of genetic determinism; too apt to believe that social stereotypes (especially about sex and inter-communal relations) reflected typical patterns of actual choice over deep time; too ready to assume a simple and direct relationship between individual profile and collective arrangements. But all that said, surely there is some relationship between the profiles of individual agents—especially stable, typical profiles—and feasible social environments? Surely it cannot be just a mistake to criticize some moral and social views as utopian, and to argue that normative systems have to be sensitive both to human cognitive limitations and to the fact that few of us are saints?

For example, standard utilitarian theories of moral action are routinely criticized for being 'too demanding': given plausible empirical assumptions about the benefits of even modest resource flows from rich countries to poor ones, they seem to require that any physically and cognitively competent agent in a first-world country should spend every day working at the highest-paying job available, and send all of that income above minimal subsistence to the most efficient international charities. For every hour not worked literally costs lives. To reject this as utopian depends not just on our views of the function of norms in human social life but also on our views of individual agents and their capacities and motivations. If most of us could not imagine a more desirable life than one of service to strangers, simple utilitarianism would tell us that our inclinations, happily, coincide with our obligations. No problem of excess moral demand there.

The point is not confined to the domain of normative theorizing. There must be some relationship between individual profile and possible social arrangement. To recycle an example I have used before: imagine that our evolved sexuality was quite different. In

---

[11] For a particularly (though unintentionally) amusing example, see Levi (1984). The poster example of this somewhat conservative agenda is Wilson (1978). For an example close to self-parody, see Palmer and Thornhill (2001).

particular, suppose that human females had cycles somewhat like those of female chimp: when in season, showing an unmistakable morphological signal of very strong sexual interest, modulated by some preferences amongst potential partners; at other times, being physiologically and psychologically unavailable. While it is hard to know exactly how human social worlds would be different under that varied sexual regime, surely they would be different.[12] The same point can be made with more mundane examples. Eleanor Ostrom is famous for her work on collective action, identifying the circumstances in which collective action problems can be solved and the circumstances in which they are much less tractable (Ostrom 1998). Tragedies of the commons can often be avoided if there are relatively few agents involved, if the decision-making environment is relatively transparent, and if the costs of cooperating and the common pool benefits from successful cooperation are commensurable. Cooperation is less apt to be stable if I am unsure whether the fish I withdraw from common pool benefits are really as valuable as the ducks my neighbour harvests. Likewise, the more difficult it is for us to track each other's actions, the less stable cooperation will be. The point here is that information transparency and resource commensurability depends on individual cognitive capacity and individual motivations. This case generalizes. The conditions under which collective action is stable are just particular cases of a more general phenomenon. Models of cooperation and cooperation failure make substantial assumptions about the capacities and motivations of agents.

Consider, for example, the existence of markets. Markets have made a huge difference to human lives, and their cognitive and motivational prerequisites are not trivial. They require not just sophisticated cognitive capacities but sensitivity to long-term outcomes, hence planning and impulse control. They require significant amounts of mutual trust (Seabright 2010). So the existence of markets, and the ability of economists to explain and predict their functioning, depend on market dynamics reflecting some basic and typical features of the individual traders. Economists explain market behaviour quite well on the assumption that in market interactions, most agents are motivated by economic considerations, and most agents can track and calculate the costs and benefits of their own options reasonably well. These are quite distinctively human traits. There have been some attempts to induce market behaviour in other primates, but these studies report a very cut-down version of a human market, both in the range of options in play and in the timeframe of decision and payoffs (Chen et al. 2006).

Hauser (2009) suggests an important twist on the idea that individual profile constrains social organization. Think of social worlds by analogy to theoretical morphospaces. Evolutionary morphologists have understood shell structure (most famously) in terms of a three-dimensional space, specifying three different axes of shell growth (Raup 1967). It turns out that real shells occupy a quite restricted range of that space of possible shell shapes, and that identifies a gap in our understanding. What is the problem with

---

[12] Le Guin (1969) is a bold attempt to imagine a social world consequent on even more fundamental differences in human sexuality—a world in which sex is not stable.

the missing shells? Hauser suggests exporting this research strategy to the evolutionary social sciences: specify the space of apparently possible societies, and identify those regions that have never been occupied. Hauser expects the innate structure of individual minds to impose significant constraints on humanly possible forms of social life. There is something intriguing about this idea, for 'gaps in nature' are indeed sometimes phenomena needing explanation in their own right. Few of the great clades of animal life have succeeded in establishing on land. Why? It is surely surprising, for example, that molluscs have not invaded the land, and that there have never been giant forest octopuses, swinging through the night forest, exploiting their multiple arms, their wonderfully protean body shapes, and their superb camouflage to snatch terrified monkeys from their perches. However, while a gap can be a phenomenon in itself, shell morphospace is an exception rather than the rule. It is unusually tractable, in that shell morphology can largely be captured through a small number of uncontroversially objective dimensions.[13] That is very unlikely to be true of human social worlds.

So there is something to be said for the idea that a theory of human nature has a role to play in understanding feasibility, and in understanding the diversity of human social life. However, that idea faces serious challenges. First, even if there is a relationship between individual profile and collective organization, it is indirect, and social possibilities may be unrealized for many reasons. One central problem in palaeoanthropology is to explain the stability of human cooperation—of the social contract—as human social worlds became larger and more complex near the Pleistocene–Holocene transition. Cooperation in small social worlds can be sustained by face-to-face mechanisms of affiliation and trust, and by the fact that in small social environments, reputation is both very important and very reliably tracked. It is hard to see how those mechanisms could be effective as the social world expands; so our knowledge of human nature generates an explanatory puzzle at the level of social organization. We know that people are often willing to cooperate for common benefit, but they are not saints. They notice if they are cooperating and others are not, and withdraw cooperation in those circumstances. They care about their reputation, and they track and respond to the reputations of others. Given all this, we would expect people to be cautious and untrusting in larger social environments in which they are interacting with near-strangers. And yet in many cases, collective action continued and expanded (Seabright 2010; Sterelny 2013a, 2013b). The predicted constraint seems not to bite.

There are more fundamental problems. The attributes of individual agents shape and constrain potential human social environments, and not just in broad and obvious ways.[14] But that truth is not well expressed by saying that human nature constrains and

---

[13] See Maclaurin and Sterelny (2008: ch. 4) for a discussion of both the promise and the problems of morphospaces and their kin.

[14] Obviously, all societies must make some provision for the extraction and distribution of resources; have some systematic provision of care for infants and children; have some controls on violence; and the like. This is a completely trivial and uncontroversial sense in which individual biology constrains social worlds.

shapes humanly possible social worlds. For one thing, individual variation within groups might well be important for the stability of some social environments: humans vary in both talents and personalities, and in complex societies, with division of labour and very different social roles,[15] that variation might well be functionally important. Both formal models and the experimental games of behavioural economists suggest that cooperation is maximized if some fraction of the group are moralistic punishers (Boyd and Richerson 1992; Bowles and Gintis 2011); agents prepared to punish free-riding at some cost to themselves, even if they have not themselves suffered from the free-riding, and will not benefit from future prosocial behaviour from the target.[16] So in thinking about the relationship between individual attributes and collective possibility, it is very important *not* to idealize to a uniform population of the typical human (or perhaps the typical male, the typical female). Even more important is the direction of causation. Social environments shape individual attributes just as individual attributes shape social environments, and this reciprocal influence takes place over many time scales.[17] Literacy, for example, transforms the mind of the literate (Heyes 2012). I have argued in other work (Sterelny 2003, 2010, 2012) that core cognitive competences depend on environmental scaffolding, and on the social organization of the learning environment of young humans. We have far richer capacities for quantitative reasoning than other primates, and we use an enormous array of material symbols. These depend on the collective invention and transmission of representational systems (like the numerals). I remarked above that the existence of markets depends on stable preference structures and the capacity to resist (albeit not perfectly) short-term temptation over long-term gain. Don Ross (2005, 2006) has argued persuasively that this stability depends on environmental support; social environments both depend on and reinforce this capacity to stay on track. Our approximating rational agency is not an intrinsic characteristic of human agents—one that makes possible a social environment that includes markets. Rather, we live in social environments that make the future salient, and which scaffold and encourage future-oriented action. These social environments partially explain our approximate rationality.

## 6.5 Norms and Natures

In the last three sections, I have agreed that there is a defensible account of human nature, but suggested that it is bland and uninformative. Explaining our nature is not a crucial explanatory project; rather, it is the aggregate result of a set of separate, though

---

[15] This may well be quite ancient (Ofek 2001).

[16] As far as I know, there have been no studies on whether there are costs if a group consists entirely of moralistic punishers. But surely there must be costs, as such agents are not conflict-averse in the face of norm violation, and in real social environments there will be conflicting viewpoints on whether a norm has been violated. So it is easy to see how such a group could slip into punishment wars, with one agent responding with punishment to perceived unfairness in another's punishment. This phenomenon—known as antisocial punishment—is known in the experimental literature.

[17] John Dupré (2002: ch. 7) presses this direction of explanation point.

loosely related, projects. Nor is it a critical resource in explaining the large and obvious differences between human social life and the lives of our closest living relatives. Could it be a crucial resource for normative theory? Perhaps with a few exceptions, the folk, I conjecture, value human lives more than other lives. Legal rights certainly accrue to humans but not animals, and this probably reflects folk moral opinion too. To the best of my knowledge (and, more importantly, to the best of Google's knowledge), no one has carried out extensive cross-cultural trolley problem surveys in which an agent has to choose between saving the life of one (or more) humans and larger numbers of non-human animals (ducks, dogs, pandas, chimps). But I suspect that untutored intuition would permit, or perhaps require, an agent to divert a trolley from a vulnerable child, even at the cost of its travelling through a flock of ducks; likewise, an agent would be morally permitted to push a large pig onto the tracks in front of the escaped trolley, slowing it enough to enable a child to escape.

Perhaps this intuition is an expression—legitimate or illegitimate—of our partiality in favour of a fellow member of our own species. But perhaps a theory of human nature would show a relevant difference between the human child and a flock of ducks—one that would make it legitimate for any rational agent to sacrifice the ducks' interests in favour of the child. The cluster-like character of the banal concept does not look problematic in this context. The human/duck difference could be legitimized by the human cognitive suite: a cluster of traits whose precise character and interrelations change over time, though all have some significance. These might include: a sense of oneself and others as intentional agents; introspective self-consciousness; episodic memory and mental time-travel; empathy and prosocial emotions; responsiveness to norms; language and the capacity for abstract thought. None needs be necessary nor sufficient in itself. But it seems very likely that across a whole spectrum of traits like these, an average human will come out with a very different profile from that of a typical Australian Black Duck.

Peter Singer (1993, 2009) and his intellectual allies have argued against the idea that an appeal to human nature can legitimize these intuitive normative responses. In a rather curious echo of David Hull, Singer relies on the fact that these phenotypic traits, either singly or as a cluster, are not strictly universal. Yet the supposed difference in moral standing is, supposedly, strictly universal. Untutored intuition presumably permits diverting the trolleybus through the flock of ducks, even if the person so saved is permanently brain-damaged, amnesiac, or has declined into severe dementia. Nor are such humans to be eaten (except, perhaps, in lifeboat choices). If our distinctive mental and social capacities underwrite and justify our distinctive normative standing, that standing should extend only to those members of our species who have those capacities (or perhaps: have or will develop them). Since normative standing is not taken to be restricted in this way, the human cognitive suite does not legitimize a Cartesian divide in moral standing between humans and other animals.[18]

---

[18] Singer's views do not commit him to thinking that duck and human life are fully morally equivalent; he can and does allow that in a straight forced choice, one can legitimately choose a human.

However, we can resist Singer's *modus tollens*. He has in mind the wrong picture of an explanatory relationship between normative and natural facts. Philip Kitcher, David Gautier, and the whole neo-contractarian tradition argue that norms are social technologies (Gauthier 1987; Kitcher 2011; Sterelny 2014). They make possible, or at least much easier, mutually profitable and relatively peaceful cooperation. Humans depend on one another, usually without any problem, even though in recent centuries this dependence is often on complete strangers. As Paul Seabright has emphasized, in doing so we ultimately rely on one another's respect for norms. It is true that this reliance has become somewhat more secure through the existence of institutional sanctions for gross violations of those expectations. But those sanctions themselves rely on actual respect for norms and the general expectation of that respect (Seabright 2006, 2010).

On this neo-contractualist understanding of human norms and their basis, a moral judgement about (say) whom to save and whom to sacrifice is not a report of objective reality in quite the same sense that a specification of a mallard's haemoglobin is an objective report on independent reality. But norms are not arbitrary: social mechanisms that support and enhance cooperation and collective action are constrained by facts about human agents and the societies those agents build (Sterelny and Fraser 2017). One of those constraints is tractability. As David Lewis remarked in discussing norms of tolerance in mixed societies, it is necessary 'to keep it simple, stupid' (Lewis 2000: 187). The normative practices of a culture need to be intelligible and memorable not just to adults; it must be possible to transmit them with reasonable fidelity to the next generation. They must be intelligible and then credible to the young. For human cultures are multi-generational ensembles of the aged, the ageing, prime adults, sub-adults, juveniles, infants. As with language, no doubt these ensembles have some degree of intergenerational variation. But it would obviously be cause for serious conflict if those differences became extensive and important. This does occasionally happen, of course, but it is not a routine feature of intergenerational turnover. Hence the social role of norms depends on fairly faithful transmission. For norms to lubricate cooperation, the incoming generation needs to understand and largely accept the normative practices of their seniors.

On this view of norms, it is no surprise that there is a marked difference in the norms regulating interactions with (a) in-group members; (b) potential in-group members and/or members of other groups with whom there is a potential peace dividend; and (c) agents with whom active cooperation is off the table. These differences are not arbitrary, but their justification does not depend on the intrinsic character of the agents in question. Moreover, given the design case in favour of explicit, simple, unambiguous norms and the fact that in most cases there is a gradient between full, paradigmatically functional agents in the community—agents with the full cognitive suite—and those that are variously a few cans short of a six-pack,[19] we might expect in-group norms to

---

[19] That said, humans are good at culturally manufacturing boundaries, which can then be recruited to conditionalize permissions and prohibitions, making them more targeted. Sex, initiation ceremonies, clan groups, moieties, and kinship systems can all impose sharp qualitative distinctions.

be generous, lumping marginal agents with full agents. When we can afford it,[20] the cracked are treated as if they had full standing—a practice probably made easier to accept by the realization that we are all just a moment or two of bad luck from being cracked ourselves.

In short, a wholly naturalistic conception of normative facts does not need a theory of human nature. Rather, it is grounded in facts about human social environments, and facts about cooperation promoters and conflict flashpoints in those environments. Of course, there is an intimate relationship between social environment and individual profile. But as discussed in the last section, the causal arrow between individual and collective character runs in both directions.

Machery and Godfrey-Smith are right: we can reject intrinsic essentialism about species in general, and our species in particular, and retain a concept of human nature. But it is a descriptive, 'field guide' concept: it does no explanatory work and is a somewhat arbitrary, list-like conception. Do we need it?

# References

Bowles, S., and Gintis, H. (2011). *A Cooperative Species: Human Reciprocity and Its Evolution*. Princeton, NJ: Princeton University Press.
Boyd, R., and Richerson, P. (1992). 'Punishment Allows the Evolution of Cooperation (or Anything Else) in Sizable Groups.' *Ethology and Sociobiology* 13: 171–95.
Boyd, R., and Richerson, P. (2013). 'Rethinking Paleoanthropology: A World Queerer Than We Supposed.' In G. Hatfield (ed.), *Evolution of Mind, Brain, and Culture*, 263–302. Philadelphia: Penn Museum Press.
Boyd, R., Richerson, P., and Henrich, J. (2011). 'The Cultural Niche: Why Social Learning Is Essential for Human Adaptation.' *Proceedings of the National Academy of Science* 108: 10918–25.
Burkart, J., Hrdy, S. B., and van Schaik, C. (2009). 'Cooperative Breeding and Human Cognitive Evolution.' *Evolutionary Anthropology* 18: 175–86.
Chen, M. K., Lakshminaryanan, V., and Santos, L. R. (2006). 'The Evolution of Our Preferences: Evidence from Capuchin Monkey Trading Behavior.' *Journal of Political Economy* 114: 517–37.
Devitt, M. (2008). 'Resurrecting Biological Essentialism.' *Philosophy of Science* 75: 344–82.
Devitt, M. (2010). 'Species Have (Partly) Intrinsic Essences.' *Philosophy of Science* 77: 648–61.
Dominguez, M. H., and Rakic, P. (2009). 'The Importance of Being Human.' *Nature* 462: 169–70.
Dupré, J. (2001). *Human Nature and the Limits of Science*. Oxford: Oxford University Press.
Dupré, J. (2002). *Humans and Other Animals*. Oxford: Clarendon Press.

---

[20] Only wealthy societies can afford to support the permanently helpless; those closer to the bone have variously locally permitted forms of termination; see Hrdy (1999) for an extended discussion of the infanticide of babies and infants with poor prospects of developing into fully competent adults. In poorer societies, this is tolerated and sometimes enforced.

Ereshefsky, M. (2010). 'What's Wrong with the New Biological Essentialism?' *Philosophy of Science* 77: 674–85.
Ereshefsky, M. (2014). 'Species, Historicity, and Path Dependency.' *Philosophy of Science* 81: 714–26.
Gauthier, D. (1987). *Morals by Agreement*. New York: Oxford University Press.
Godfrey-Smith, P. (1993). 'Functions: Consensus without Unity.' *Pacific Philosophical Quarterly* 74: 196–208.
Godfrey-Smith, P. (1994). 'A Modern History Theory of Functions.' *Noûs* 28: 344–62.
Godfrey-Smith, P. (2014). *Philosophy of Biology*. Princeton, NJ: Princeton University Press.
Gould, S. J., and Eldredge, N. (1993). 'Punctuated Equilibrium Comes of Age.' *Nature* 366: 223–7.
Hauser, M. (2009). 'The Possibility of Impossible Cultures.' *Nature* 460: 190–96.
Hawkes, K., O'Connell, J. F., Blurton Jones, N. G., Alvarez, H., and Charnov, E. (1998). 'Grandmothering, Menopause, and the Evolution of Human Life Histories.' *Proceedings of the National Academy of Science USA* 95: 1336–9.
Henrich, J. (2004). 'Demography and Cultural Evolution: Why Adaptive Cultural Processes Produced Maladaptive Losses in Tasmania.' *American Antiquity* 69: 197–221.
Henrich, J. (2006). 'Cooperation, Punishment, and the Evolution of Human Institutions.' *Science* 312: 60–1.
Heyes, C. (2012). 'Grist and Mills: On the Cultural Origins of Cultural Learning.' *Philosophical Transactions of the Royal Society B* 367: 2181–91.
Hrdy, S. B. (1999). *Mother Nature: A History of Mothers, Infants, and Natural Selection*. New York: Pantheon Books.
Hrdy, S. B. (2009). *Mothers and Others: The Evolutionary Origins of Mutual Understanding*. Cambridge, Mass.: Harvard University Press.
Hull, D. (1986). 'On Human Nature.' *Proceedings of the Philosophy of Science Association* 1: 3–13.
Kitcher, P. (1985). *Vaulting Ambition: Sociobiology and the Quest for Human Nature*. Cambridge, Mass.: MIT Press.
Kitcher, P. (2011). *The Ethical Project*. Cambridge, Mass.: Harvard University Press.
Klein, R., and Edgar, B. (2002). *The Dawn of Human Culture*. New York: Wiley.
Le Guin, U. (1969). *The Left Hand of Darkness*. New York: Ace Books.
Levin, M. (1984). 'Why Homosexuality Is Abnormal.' *The Monist* 67: 251–83.
Lewens, T. (2012a). 'Human Nature: The Very Idea.' *Philosophy and Technology* 25: 459–74.
Lewens, T. (2012b). 'Species, Essence, and Explanation.' *Studies in History and Philosophy of Biological and Biomedical Sciences* 43: 751–9.
Lewis, D. (2000). 'Mill and Milquetoast.' In his *Papers in Ethics and Social Philosophy*, 159–87. Cambridge: Cambridge University Press.
Machery, E. (2008). 'A Plea for Human Nature.' *Philosophical Psychology* 21: 321–9.
Maclaurin, J., and Sterelny, K. (2008). *What Is Biodiversity?* Chicago: University of Chicago Press.
Mameli, M., and Bateson, P. (2011). 'An Evaluation of the Concept of Innateness.' *Philosophical Transactions of the Royal Society B* 366: 436–43.
Mithen, S. (1996). *The Prehistory of the Mind*. London: Phoenix Books.
Mithen, S. (2009). 'Holistic Communication and the Coevolution of Language and Music: Resurrecting an Old Idea.' In R. Botha and C. Knight (ed.), *The Prehistory of Language*, 58–76. Oxford: Oxford University Press.

Mithen, S. (2013). 'The Cathedral Model for the Evolution of Human Cognition.' In G. Hatfield and H. Pittman (ed.), *Evolution of Mind, Brain, and Culture*, 217–34. Philadelphia: University of Pennsylvania Press.
Morley, I. (2013). *The Prehistory of Music*. Oxford: Oxford University Press.
Ofek, H. (2001). *Second Nature: Economic Origins of Human Evolution*. Cambridge: Cambridge University Press.
Okasha, S. (2002). 'Darwinian Metaphysics: Species and the Question of Essentialism.' *Synthese* 131: 191–213.
Ostrom, E. (1998). 'A Behavioral Approach to the Rational Choice Theory of Collective Action.' *American Political Science Review* 92: 1–22.
Oyama, S., Griffiths, P. E., and Gray, R. (2001). *Cycles of Contingency: Developmental Systems and Evolution*. Cambridge, Mass.: MIT Press.
Palmer, C., and Thornhill, R. (2001). *A Natural History of Rape: Biological Bases of Sexual Coercion*. Cambridge, Mass.: MIT Press.
Raff, R. (1996). *The Shape of Life: Genes, Development, and the Evolution of Animal Form*. Chicago: Chicago University Press.
Raup, D. (1967). 'Geometric Analysis of Shell Coiling: Coiling in Ammonoids.' *Journal of Paleontology* 41(1): 43–65.
Ross, D. (2005). *Economic Theory and Cognitive Science: Microexplanation*. Cambridge, Mass.: MIT Press.
Ross, D. (2006). 'The Economic and Evolutionary Basis of Selves.' *Cognitive Systems Research* 7: 246–58.
Seabright, P. (2006). 'The Evolution of Fairness Norms: An Essay on Ken Binmore's *Natural Justice*.' *Politics, Philosophy, and Economics* 5: 33–50.
Seabright, P. (2010). *The Company of Strangers: A Natural History of Economic Life*. Princeton, NJ: Princeton University Press.
Singer, P. (1993). *Practical Ethics: How Are We to Live?* Cambridge: Cambridge University Press.
Singer, P. (2009). 'Speciesism and Moral Status.' *Metaphilosophy* 40: 567–81.
Sterelny, K. (2003). *Thought in a Hostile World*. Chichester: Wiley.
Sterelny, K. (2007). 'Social Intelligence, Human Intelligence, and Niche Construction.' *Proceedings of the Royal Society London B* 362: 719–30.
Sterelny, K. (2010). 'Minds: Extended or Scaffolded?' *Phenomenology and the Cognitive Science* 9: 465–81.
Sterelny, K. (2012). *The Evolved Apprentice*. Cambridge, Mass.: MIT Press.
Sterelny, K. (2013a). 'Co-operation in a Complex World.' *Biological Theory* 7: 358–67.
Sterelny, K. (2013b). 'Life in Interesting Times: Co-operation and Collective Action in the Holocene.' In B. Calcott, B. Fraser, R. Joyce, and K. Sterelny (eds), *Cooperation and Its Evolution*, 89–108. Cambridge, Mass.: MIT Press.
Sterelny, K. (2014). 'A Paleolithic Reciprocation Crisis: Symbols, Signals, and Norms.' *Biological Theory* 9: 65–77.
Sterelny, K., and Fraser, B. (2017). 'The Prospects for Moral Realism.' *British Journal for the Philosophy of Science* 68: 981–1006.
Tomasello, M. (1999). *The Cultural Origins of Human Cognition*. Cambridge, Mass.: Harvard University Press.

Tomasello, M. (2014). *A Natural History of Human Thinking.* Cambridge, Mass.: Harvard University Press.
Wilson, E. O. (1978). *On Human Nature.* Toronto: Bantam Books.
Wrangham, R. (2009). *Catching Fire: How Cooking Made Us Human.* London: Profile Books.
Wynn, T., and Coolidge, F. (2012). *How to Think Like a Neanderthal.* New York: Oxford University Press.

# 7

# The Social Construction of Human Nature

*Kevin N. Laland and Gillian R. Brown*

## 7.1 Introduction

Scientific terms have to earn their keep.* What is the job that the term 'human nature' is designed to do? Here we discuss three possibilities: (1) To distinguish what is deemed 'biological' from that which is deemed 'cultural' or 'environmental'. (2) To characterize the defining features of humanity, thereby allowing us to be distinguished from other species. (3) To characterize what is universal about humanity, or typical of it, because of our 'evolved biological heritage'. Below, we argue that the first notion of 'human nature' is long discredited, that the second is tenable but contributes little, and that the third is tenable but both misleading and counterproductive. We suggest that current conceptions of human nature frequently endeavour to fulfil several functions simultaneously (e.g. describing humanity, explaining our origins, distinguishing humans from other species), which, since these are not mutually compatible, results in the term providing sub-optimal solutions to all functions. We conclude that the term is more trouble than it is worth and should be abandoned. Instead, we advocate a more inclusive understanding of the human condition, one that provides a more complete causal-explanatory account of its origins, and that embraces variation as much as commonality. Such an understanding can be specified as the product of internal and external constructive processes operating over both developmental and evolutionary timescales.

## 7.2 Three Beleaguered Conceptions of Human Nature

### 7.2.1 Why it is not possible to distinguish what is 'biological' from what is environmental/cultural

The history of accounts of human behavioural development is characterized by a constant swinging of the pendulum to favour explanations for behaviour in terms of

---

\* Research supported in part by a grant to Kevin N. Laland from the John Templeton Foundation. We are grateful to John Odling-Smee and Tim Lewens for helpful comments on an earlier draft and Pete Richerson for valuable discussion.

nature or nurture (Boakes 1984; Bateson and Martin 1999; Laland and Brown 2011). At the extreme ends of these debates, an emphasis on 'nature' typically stresses the importance of internal cues, such as genes, and behaviour is described as being underpinned by 'innate' or 'instinctive' mechanisms. Development is portrayed as highly canalized, or channelled in particular directions. Environmental factors, learning, and culture are portrayed as relatively unimportant. In contrast, an emphasis on 'nurture' places greater weight on the importance of external cues, such as environmental and social factors, and behaviour is said to be highly dependent on learning and culture. Development is portrayed as extremely flexible, and genes, 'innate' behaviour, and 'instincts' are deemed relatively unimportant. These debates are well rehearsed and we will not elaborate greatly on them here, other than to say that the polarization is highly misleading (Bateson and Martin 1999; Ridley 2003; Keller 2010). There is no nature/nurture dichotomy, since human development, including our behaviour and cognition, is always reliant on both 'nature' and 'nurture' (Bateson and Martin 1999; Ridley 2003; Keller 2010).

Historically, researchers (e.g. Lorenz 1965) have tried to separate our 'biology' from environmental (including cultural) influences, which ultimately proved impossible (Bateson and Martin 1999). Comparative psychologist Daniel Lehrman famously critiqued the ethologists' concept of 'innate' and 'instinctive' behaviour, drawing on his experimental studies of courtship in doves (Lehrman 1953: 1964). He showed how courtship behaviour was not specified by an 'innate genetic programme', but rather involved an intricate interplay of internal (e.g. hormonal) and external (e.g. social stimuli) variables. For instance, the male dove's courtship triggers a change in the female dove's hormones, exciting gonadotropin secretion, which initiates sexual receptivity and nest-building behaviour (Lehrman 1964; Slater 1978). Numerous examples of the interplay between variables that have been characterized as being part of 'nature' and 'nurture' now exist. For instance, birdsong learning and production are reliant on the development of the 'song system', a well-mapped neural circuitry of brain nuclei and their projections within the songbird brain, which emerges in early ontogeny and subsequently matures in functionality as a result of a constant interplay between internal and external causal influences (Bolhuis and Gahr 2006; Catchpole and Slater 2008; Zeigler and Marler 2008).

The idea that 'innate' or 'instinctive' behaviour patterns can be defined as evolved behaviour that is present at birth and remains unaffected by experience has been resoundingly rejected (Bateson and Mameli 2007; Mameli and Bateson 2011), largely based on empirical evidence of how behavioural development actually unfolds. For instance, Gilbert Gottlieb (1971) showed that a newly hatched duckling's preference for the maternal calls of its own species (hitherto regarded as 'instinctive') was affected by hearing its own vocalizations in the egg. This preference is a characteristic that is present at birth but not unlearned. Attempts to identify instincts through isolating developing animals from all relevant experience failed because it is impossible to separate an animal from itself, and hence the animal will always have some experience that affects the character's development (Bateson and Martin 1999). Recognition of this fact left the concept of 'instinct' virtually impossible to test in practice. While establishing experimentally that a behaviour is not learned requires the exclusion of all opportunities for learning, this is difficult in practice because animals can acquire relevant experience in numerous ways. For instance, cats can acquire and improve their adult predatory skills by catching live prey when young, by playing at catching, by watching their mother catch prey, by playing with siblings, by catching prey as an adult, and so on (Bateson and Martin 1999).

In a well-crafted critique, Lehrman (1953) dismissed the concept of 'innate' behaviour, first, because organisms never develop in complete isolation from their environment and hence one could never know that a behaviour pattern was uninfluenced by external events, and second, 'innate' was defined in terms of excluding what is learned, rendering the concept unusable. Lehrman also pointed out it does not follow that behaviour patterns that are universal to a particular species are necessarily innate, as the apparent ubiquity of a trait does not indicate anything about its mode of development. For instance, human beings across all continents of the world warm themselves by wearing clothes and through the use of fire or other heating technology, but only an extreme nativist would claim that these behaviour patterns are 'innate'. Earlier, Theodore Schneirla had argued that the relative importance of 'innate' and 'acquired' effects on behaviour could not be separated, and that an individual's development is a complex interaction of genetic information, the developing organism, and its environment (Hinde 1982). In spite of these numerous critiques, the false distinction between innate and learned behaviour permeates discussions about behaviour and cognition to the present day (for critiques, see Bateson and Martin 1999; Keller 2010; Ridley 2003).

The false dichotomy of genes and environment is highlighted when we realize that many of the genes expressed in our body are themselves environmentally acquired. The human microbiome is the community of bacteria, archaea, fungi, and protozoa that cohabit our body cavities, surfaces, and tissues. These symbionts are acquired in part by vertical transmission from our mothers (e.g. bacteria within the egg cells) and in part from the external environment. The microbiome is now understood to play crucial roles in normal development, as well as in nutrient acquisition, metabolism, immune function, and behaviour. About one third of the metabolites in our blood are derived from bacteria. We have 22,000 genes of our own, but in our bodies are more than 3 million genes belonging to our symbionts, which play important roles in our development, which is a multi-species project. In mammals, symbionts have been shown to be responsible for the development of the gut capillaries and even for brain development. Human microsymbionts play a major role in the development of immune and metabolic systems and brain function (Cryan and Dinan 2012; Chiu and Gilbert 2015).

The futility of trying to separate genes from environment, and the inappropriateness of attempting to characterize only those processes that involve genes as being 'biological', is clearly apparent when the action of genes is examined. Genes do not exert their influence on development independently of the environment, just as 'nature' is not independent of 'nurture'. Both internal and external environmental factors regulate gene expression (Ridley 2003; Gilbert 2003). For instance, learning cannot occur without gene expression, as learning involves the transcription of genes coding for proteins that alter the properties of neurons and synapses. These genes include 'immediate early genes' that are transcribed very rapidly, usually within minutes after the stimulation of a cell. As an example, elevated levels of expression of the immediate early gene ZENK are associated with song learning in male canaries (Jarvis and Nottebohm 1997). Likewise, during behavioural imprinting, through which young birds learn to attach to their mothers, an increase in the expression of the immediate early gene *c-fos* occurs within approximately one hour of imprinting training, which leads to a cascade of pre- and post-synaptic changes in the brain regions where the memory is stored (Horn 2004). The same interactions are characteristic of human learning (Daw and Shohamy 2008). In fact, numerous environmental influences with a lasting influence on the phenotype influence gene expression. For example, studies have shown that both short- and long-term practitioners of meditation and yoga show significantly altered gene expression profiles: genes associated with stress responses are consistently up- or down- regulated, providing a cellular mechanism for the long-known health benefits of these practices (Dusek et al. 2008).

More generally, rather than being the context in which development takes place—the 'theatre' in which a genetically orchestrated performance ensues—the environment is *always* actively involved in shaping the developmental phenotype. Developmental plasticity is now recognized by evolutionary and developmental biologists to be ubiquitous at all levels of biological organization (West-Eberhard 2003; Gilbert and Epel 2009; Sultan 2015).

Several strands of evolutionary thinking—including evolutionary developmental biology, ecological developmental biology, developmental systems theory, and niche construction theory—encompass the idea that organisms are reciprocally caused by endless cycles of constructive processes, which inextricably interweave internal and external factors throughout ontogeny (Oyama 1985; Odling-Smee et al. 2003; Noble 2006; Gilbert and Epel 2009; Keller 2010). From this perspective, separating the organism from its environment is highly problematic, as organisms not only respond plastically to environmental conditions, but also construct and regulate important components of their environment, thereby influencing both their own development and their own evolution (Lewontin 1983, 2000; Odling-Smee et al. 2003; Sultan 2015). The webs, mounds, and dams that are produced by spiders, termites, and beavers, as well as the urban environments built by human beings, are examples of 'niche construction' (Odling-Smee et al. 2003). The term 'nature' is not easily accommodated within this framework. The spider's web, for example, is a reflection of 'internal' processes that help account for its manufacture, but is also a critical feature of the spider's environment (Herberstein 2011). A key characteristic of the niche-construction perspective is the emphasis placed on the reciprocal causal processes that act alongside natural selection during the evolution of traits (Odling-Smee et al. 2003; Scott-Phillips et al. 2014), and this same emphasis is central to the emerging extended evolutionary synthesis (Laland et al. 2014, 2015).

The structures that animals build often dampen variability in environmental conditions, with the result that niche construction can maintain selection pressures and preserve the adaptiveness of behavior (Laland and Brown 2006). Odling-Smee et al. (2003) define 'counteractive niche construction' as occurring when organisms either perturb their environments or relocate in space to neutralize some prior change in selection pressures. The ability of human beings to engage in counteractive niche construction is amplified by their capacity for culture. Like the food-storing mammal or the wasp that cools its nest with droplets of water, human beings ensure the availability of food by tracking game, farming, and storing food, and we control temperature by building fires and shelters. In principle, modern refrigerators and air conditioning are little different from the thermoregulatory and gas-exchange activities exhibited by termites and ants. Such niche construction can negate a modified or fluctuating selection pressure, thereby reducing selection (Odling-Smee et al. 2003). Thus, while human-built environments might differ substantially from the African savannah or forest context that were inhabited by our hominin ancestors, many selection pressures could be broadly similar, as our constructions are likely to be suited to our bodies and their needs (Laland and Brown 2006).

Like other organisms, humans construct nursery environments that are safe and benign places in which their offspring can develop. This ecological legacy is known as 'ecological inheritance' (Odling-Smee et al. 1996, 2003). Ecological inheritance constitutes a constructed developmental niche that shapes the development of descendant organisms. This means that niche construction constitutes an important way in which environmental factors become incorporated into normal development, as well as a means of enabling some developmental factors of environmental origin to become as dependable as genomic factors. For instance, baby birds reliably spend the early part of their lives in a protected, benevolent, warm, and temperature-regulated environment, in the form of the nest that their parents build for them. Likewise, the larvae of hundreds of thousands of species of insects reliably hatch in the vicinity of a suitable food source through the reliability of their mother's host choice for egg laying (Odling-Smee et al. 2003). To quote Badyaev and Uller (2009: 1172):

Species-specific phenotypes are as much a product of species-specific environments of development as they are of species-specific genotypes, and parental effects play a crucial role in constructing such environments. Some of these parental effects, such as maternally derived RNAs, organelles, cytoplasmic gradients and symbionts, can be transferred directly to influence developmental variation in offspring, and some parental effects in this category provide a developmental template and resources for early embryonic morphogenesis.

In other words, the constancy or stability of characters across generations, generally described as 'heredity', in part results from the consistent and repeated reconstruction of developmental environments by niche-constructing organisms (Badyaev and Uller 2009).

Humans are no different. Human beings actively construct physical, developmental, and epistemological environments for their offspring. As parents, we control the temperature, humidity, food and water availability, and safety of the environments in which our children grow up. We actively create learning opportunities for our children, ensuring that they are exposed to stimulating environmental conditions in which their brains will develop to extract, process, and store relevant information (Stotz 2010; Sterelny 2012). We rear our children in an environment rich in language-learning opportunities. We speak to our children in 'infant-directed speech', we teach them verbal labels and syntax, we correct their linguistic errors, we reinforce every step through encouragement.

In sum, no human trait can be identified that is exclusively fashioned by genetic factors, and all human traits are the product of complex gene-environment interactions. There is no nature/nurture dichotomy. Genes take their cues from the environment, while learning relies on gene expression. The development of an individual animal cannot be divided into components that are either learned or innate, since all development is influenced by both genes and the environment. The notion of 'human nature' as capturing that which is solely, or mostly, influenced by genes is long discredited. Were researchers to document a constellation of reliably developing human capacities that are more or less ubiquitous in the species, and whose development seems to be well buffered against broad environmental fluctuations, then there would be no reason to attribute such traits to 'nature' as opposed to 'nurture', 'culture', 'learning', or 'environment'.

### 7.2.2 *The defining features of humanity*

The second conception of human nature sets out to characterize the defining features of humanity, thereby allowing our species to be distinguished from other species. We suggest that this is tenable, but does little useful work, as one can specify what is distinctive about humanity equally well (in fact, probably with greater clarity) without the concept of 'human nature'. Moreover, this perspective is problematical too, as it is also essentialist, and does not sit well with the fission/fusion nature of biological reality—for instance, the recently detected interbreeding of humans with Neanderthals (Green et al. 2010) and Denisovans (Krause et al. 2010).

Human populations have variant evolutionary histories, in both space and time. If we focus on 'unique' features that separate humans from the rest, then, curiously, some of the distinctive features of our species include our culture and our (niche-constructed) environments. On this perspective, our 'nature' would equate to what has traditionally been viewed as our 'nurture'. Conversely, if it is the *set* of properties of a species that is distinctive, then we must accept that many, often most, of the individual properties that characterize a species are shared with closely related species (Hull 1986). On this view, much of human nature is shared with other animals, since many key characteristics of humanity (our imitation, social learning, tool use, extractive foraging, hunting, theory of mind, complex social lives) are probably shared with other apes.

We emphasize that the properties that allow species to be distinguished are often quite different from those seem to capture their 'essence' (Lewens 2015). For example, evolutionary biologists recently detected a large number of 'cryptic' species that ostensibly appear identical and often can only be distinguished genetically. There are more than 1,500 species of fruit flies (*Drosophila* spp.), for instance, many of which are morphologically virtually identical, and can only be identified using molecular data or through attending to very subtle differences in their courtship song or olfactory signals. The question then is whether each of these species has its own 'nature' specified by its distinctive signals, or whether we recognize that *Drosophila* species share suites of characters (flies, small size, fruit feeding, and so on). In 2001, researchers discovered that the African elephant is actually two separate and non-interbreeding species, now known as the forest and savannah elephant (Roca et al. 2001). The forest elephants have slightly thinner tusks and rounder ears, which raises the question of whether these traits are the defining properties of the 'nature' of forest elephants. A positive answer would be unsatisfactory, as the characters that shape elephant biology, and allow them to be distinguished from other animals, are precisely those that forest elephants share with savannah elephants (their large size, their trunk, their long lives, their social structure, and so on).

The key point here is that while it is perfectly possible to distinguish humans from other extant animals, the criteria that allow this are quite distinct from the criteria that capture the 'essence' of human nature. It follows that any human nature concept that tried to do both jobs must do at least one sub-optimally.

### 7.2.3 Why it is misleading to attempt to characterize what is universal about humanity in terms of our 'evolved biological heritage'

The conception of 'human nature' as an umbrella term for a package of what are claimed to be universal, evolved human characteristics has a long history within the human evolutionary behavioural sciences. Sociobiologist Edward Wilson (1994: 332–3) summarized this version of human nature:

Human beings inherit a propensity to acquire behavior and social structures, a propensity that is shared by enough people to be called human nature. The defining traits include division of labor between the sexes, bonding between kin, incest avoidance, other forms of ethical behavior, suspicion of strangers, tribalism, dominance orders within groups, male dominance overall, and territorial aggression over limiting resources. Although people have free will and the choice to turn in many directions, the channels of their psychological development are nevertheless [...] cut more deeply by the genes in certain directions than in others. While cultures vary greatly, they inevitably converge toward these traits.

This package, which closely parallels later lists of universal human traits (e.g. Brown 1991), typically equates to those traits studied and emphasized by evolutionary psychologists (e.g. Buss 1999; Pinker 2003), although evolutionary psychologists typically give them a more psychological focus. Much has been written on Wilson's (1994) claims that certain properties of human behaviour and cognition are stable, inevitable, or difficult to change, and the debates surrounding human sociobiology have been extensively documented (Laland and Brown 2011; Segerstråle 2000). Despite the criticisms of human sociobiology and evolutionary psychology, the concept of evolved psychological mechanisms that are universal and that result from a history of natural selection acting on genetic variation has persisted.

According to Tooby and Cosmides (2005: 5), the 'long-term scientific goal toward which evolutionary psychologists are working is the mapping of our universal human nature'. In recent times, evolutionary psychologists have tended to emphasize the apparent flexibility and contingency of human behaviour, which is believed to be underpinned by 'universality' (or at least typicality) at the level of psychological mechanism. This putative universal cognition is rendered compatible with diversity in human behaviour through context- or condition-dependent strategies (e.g. Gangestad and Simpson 2000). For instance, differential environmental influences are deemed to act like switches that shift the behavioural output of this universal genetic programme in an adaptive, condition-specific manner, in a manner analogous to how the buttons of a jukebox change the tunes it plays (Gangestad and Simpson 2000; Tooby and Cosmides 1992). The presence of evolved psychological mechanisms that underpin variable responses according to key environmental parameters thus appears to be compatible with the assumption that models from behavioural ecology can be equally applied to human beings and to other animal species (Cashdan 2013; Kaplan and Gangestad 2005).

We regard 'universality' as something of a red herring in this debate. That is, while we do not defend the notion of human nature as 'characterizing that which is universal or typical about humanity because of our evolved biological heritage', we do not believe that, in practice, this conception stands or falls on whether universality can be demonstrated. In our view, both sides of the debate fully recognize that humans, like all other species, are both genetically and phenotypically variable, and that this variation is manifest both in space and over time. The traits that are portrayed as human universals are not generally regarded as necessary and sufficient conditions for identifying humans, as Hull's (1986) critique of human nature seemingly implied. To our knowledge, nobody is making the argument that a person would not be a human being if he or she did not exhibit the characteristics that sociobiologists and evolutionary psychologists regard as central to human nature (the same holds for societies that fail to exhibit group-level traits commonly characterized as universal). The key point here is that for the conception of human nature defended by evolutionary psychologists, it does not really matter whether or not the so-called universal traits really are universal or whether they are merely typical (Machery 2008).

Problems arise in part because this conception of human nature is reliant on an outdated, overly gene-centric, and unidirectional model of developmental and evolutionary causation (Laland et al. 2011). The evolutionary psychologists' model of development as being reliant on genetically programmed switches (i.e. the jukebox metaphor)—determining which phenotype will be produced with each environment encountered—is a poor model of human development (Oyama et al. 2001; Noble 2006; Keller 2010). Phenotypes are not the simple causal products of the execution of genetic programmes, but rather are constructed by a reciprocally caused process that comprises both upward and downward causation (Oyama et al. 2001; Noble 2006; Keller 2010). Development is not pre-programmed; it is constructive (Laland et al. 2015). The developmental system responds flexibly to internal and external inputs, most obviously through condition-dependent gene expression, but also through physical properties of cells and tissues and 'exploratory behaviour' among microtubular, neural, muscular, and vascular systems. For example, there is no predetermined map for the distribution of blood vessels in the body; rather, the vascular system expands to regions with insufficient oxygen supply. Such exploratory processes, commonplace throughout development, are powerful agents of phenotype construction, as they enable highly diverse functional responses that need not have been pre-screened by earlier selection (West-Eberhard 2003; Kirschner and Gerhart 2005). Organisms are not built from genetic 'instructions', but rather self-assemble using a broad variety of interdependent resources. Moreover, development occurs in a self-constructed environment designed to scaffold individual development and learning (Stotz 2010; Sterelny 2012). One ramification of constructive development, then, is that functionality need not imply adaptation (i.e. need not imply a history of selection for that character). More to the point here, if this constructive model of development is correct, there can be no genetic programme underlying a universal cognition because no such programme exists. Rather, the evolutionary process has bootstrapped a series of ontogenetic exploratory and selective processes that confer enormous plasticity on human development.

A second set of problems arises from the attribution of the stability of 'universal' human traits to naturally selected human genes. Recent findings within evolutionary biology have led to a broadened conception of inheritance, and to the recognition that parent–offspring similarity does not solely result from the transmission of genes from one generation to the next, but also from the transmission of other resources, by a variety of different pathways (Badyaev and Uller 2009; Gilbert and Epel 2009; Danchin et al. 2011; Jablonka and Lamb 2014). Examples include (i) the germline transmission of epigenetic variants, such as DNA methylation and small RNAs, which can lead to stable inheritance of phenotypes in the absence of differences in structural DNA, as in the epigenetic inheritance of differences in fruit size, flowering time, and other traits in *Arabidopsis*. Inheritance can also involve (ii) parental effects such as the transfer of antibodies, hormones, symbionts, and so forth (Badyaev and Uller 2009; Gilbert and Epel 2009); (iii) cultural inheritance, defined broadly to include all forms of social transmission, which has now be shown to be prevalent in many hundreds of species of animals (Hoppitt and Laland 2013); and (iv) ecological inheritance, whereby organisms through their niche construction modify the environmental resources experienced by their descendants (Odling-Smee et al. 2003). These developmental resources enable the construction and stable reconstruction of developmental environments.

In the light of this broadened conception of inheritance, it would be unwise a priori to attribute any property of universality or even typicality to evolved human genes, or to human genetic evolution. Those human traits that are relatively stable across environments and cultures do not derive their stability solely from inherited genes, but also from diverse non-genetic inheritances, including active constructive cultural processes through which humans reliably build developmental environments for their offspring. We agree that certain aspects of behavioural development exhibit consistency and stability across space and time. For instance, human beings pass the same developmental milestones as they grow up, with most children starting to walk by 18 months of age and starting to talk by 2 years. However, simply characterizing these properties as elements of our 'human nature', with the implication that genetic inheritance plays the major role, is misleading, as the role of socially constructed developmental environments and other forms of non-genetic inheritance are also crucial. Infants consistently start walking and talking at the specified ages in part because adult humans consistently construct the right kind of developmental environment in which they can learn these skills. Moreover, growing evidence that even apparently basic cognitive processes vary greatly between populations (Henrich et al. 2010; Nisbett 2003) raises the question of how universal traits can be defined and studied (Norenzayan and Heine 2005).

There are many typical human qualities—most humans use money or barter for clothes or mobile phones, for instance—that are not immediate legacies of recent genetic evolution, although they are certainly reliant on evolved capabilities. Edouard Machery (2008: 69) acknowledges this:

The belief that water is wet is not part of human nature in spite of being common, because this belief is not the result of some evolutionary processes. Rather, people learn that water is wet.

He thereby characterizes human nature as that subset of human-typical behaviour and cognition that can be explained by direct reference to 'biological' evolutionary processes:

Only some [human] traits can be explained by reference to evolutionary processes. That is, only some of them are the object of ultimate explanations. What distinguishes these traits from the traits that are not the object of ultimate explanations is that they have an evolutionary history.

Here Machery draws on Ernst Mayr's (1961) distinction between proximate and ultimate causation. We believe that Machery's position is in line with many evolutionary psychologists in attempting to delineate those typical aspects of human behaviour that are legacies of 'biological evolution'. However, for the reasons given above, it is hard, perhaps impossible, to separate the 'biological' from the 'non-biological'. Moreover, any recourse to 'ultimate explanations' to clarify the concept of human nature amounts to the use of one dubious concept to bolster another.

The real problem with this conception of human nature is the challenge of demonstrating, in a non-arbitrary, rigorous, and scientifically useful manner, that such traits are products of 'biological' evolution, or that they have an 'ultimate explanation'. Mayr's proximate–ultimate dichotomy is itself highly problematic (Laland et al. 2011, 2012, 2013). Clear distinctions between cause, function, and evolutionary history are valuable, but these can be made without Mayr's terminology (e.g. Tinbergen 1963), which discourages full consideration of the role of development in evolution. Current use of the term 'ultimate causation' is highly confused, with some researchers reading 'ultimate' as specifying the function of a trait, others its history, and others both (Laland et al. 2013). Some researchers equate ultimate cause with a history of selection, whilst others have a broader notion that includes other population genetic processes, such as drift, mutation, and gene flow. More problematic than this ambiguity is the fact that there is no consensus over what phenomena are appropriately characterized as explained by 'proximate' and 'ultimate' causes, since the recognition of 'ultimate causes' is affected by the conceptual framework of the researcher (Laland et al. 2011, 2012, 2013). Such disagreements underlie many ongoing debates spanning the biological sciences, including debates over language, cooperation, and cultural evolution—topics squarely in the frame as aspects of 'human nature' (Laland et al. 2011, 2012, 2013).

By wrapping together function and history in the same term, Mayr and his disciples are forced to the position that the only explanation for a particular function is a history of natural selection. It would be a mistake to assume that all semblance of design arises solely from natural selection on genetic variation. We foresee many characters that exhibit design features but are not biological adaptations, including spandrels, exaptations, and products of cultural evolution, as well as the appearance of design brought about through niche construction (Laland et al. 2013).

For instance, Catholicism, the legal system, and space rockets are all products of an evolutionary process, and possess an evolutionary history that encompasses the repeated, typically gradual refinement and recombination of earlier variants. If 'ultimate explanation' is to be read as 'possesses an evolutionary history', then we maintain that this would apply equally to cultural adaptations. If space rockets are as much part of human nature as incest taboos, then perhaps this notion of human nature is too tricky to be useful. Admittedly, those contemporary academics that champion this conception of human nature are evolutionary psychologists (e.g. Pinker 2003; Tooby and Cosmides 2005), who are not typically advocates of cultural evolution (Laland and Brown 2011) and who would dispute that it affords 'ultimate' explanations (Laland et al. 2011). Such researchers might well respond by claiming that they are particularly interested in those universal traits that are the direct product of 'biological' or 'genetic' evolution—in fact, those that are 'biological' adaptations—and they might claim that it is not difficult in practice to identify these and to distinguish them from other traits. However, we are not convinced.

Recent human evolution is almost certainly strongly shaped by gene–culture coevolutionary dynamics (Laland et al. 2010; Richerson et al. 2010). Theoretical biologists have used population genetic models to demonstrate that cultural processes can have a profound effect on human evolution (Cavalli-Sforza and Feldman 1981; Boyd and Richerson 1985), and anthropologists have identified many cultural practices that modify current selection (Durham 1991; Wrangham et al. 1999; Perry et al. 2007; Cochran and Harpending 2009). These findings are supported by recent analyses of human genetic variation, which reveal that hundreds of genes have been subject to recent positive selection, often in response to human cultural activities (Voight et al. 2006; Wang et al. 2006; Sabeti et al. 2007; Nielsen et al. 2007). This means that human evolutionary history is not easily carved up into that which is biological evolution and that which is cultural evolution, since these two dynamical processes clearly interact (Laland et al. 2010; Richerson et al. 2010; Richerson 2014).

Where is the line to be drawn to delineate those 'evolved' traits that qualify as human nature from other prevalent characters? There is evidence that the human cultural practice of cooking coevolved with genes (e.g. MYH16, ENAM) that are expressed in the brain and digestive tract, and are involved in the determination of tooth size and the reduction in jaw muscle (Richards et al. 2003; Stedman et al. 2004). It might seem reasonable to regard the physical properties of human teeth, jaws, and digestive system, and perhaps the tendency to prepare food through cooking, as aspects of human nature, since they are all typical and all possess an evolutionary history. However, the morphological traits are the products of gene–culture coevolution, while the cooking is learned and culturally transmitted. One might take the line that the morphological, but not the behavioural trait, meets definitions of human nature; but that seems odd, given that cooking behaviour must have been widespread and reliably present over time in order to act as the cause of the evolutionary response in morphology, and likely preceded it in time. What is more, this line of reasoning would seem to demote the flagship example of human language from the list of traits that qualify as human nature, merely because it is learned.

Conversely, if researchers adopt the position that any human-typical trait where there is evidence of genetic evolution qualifies as an aspect of 'human nature', they run into a different class of difficulties of implementation. There is now good experimental and theoretical evidence that gene–culture coevolution generated an evolved psychology—comprising a motivation to teach, to speak, to imitate and emulate, docility, social tolerance, the ability to share the goals and intentions of others, joint attention, a tendency to produce and attend to infant-directed speech, pedagogical cuing, and much more, including extensive changes in the human brain—that is directly responsible for the human capacity for cumulative culture (Tomasello 1999; Gergely and Csibra 2005; Richerson and Boyd 2005; Tennie et al. 2009; Richerson et al. 2010; Dean et al. 2012; but see Heyes, Chapter 4 this volume). This line of reasoning seems to lead to the position that human technology, which has a history that goes back at least two million years to the earliest stone tools, is an aspect of human nature, since it coevolved with the aforementioned psychological capabilities. If that is the position, then it once again becomes germane to ask what is *not* human nature.

An evolutionary psychologist might well be frustrated by this response, since in refuting the utility of the concept of human nature, we have specified a series of typical, evolved human traits (teaching, imitation, shared attention, and so on) that might well meet their criteria for constituting human nature. However, we reiterate that we accept that human-typical evolved traits exist; we are simply not convinced that it is helpful to label these as 'human nature'. In our eyes, these traits are every bit as much the product of 'human nurture', and would remain every bit as interesting and important, if scientists were subsequently to establish that their production does not rely on the expression of recently selected human genes. Gene–culture coevolutionary research also draws attention to many additional human traits that have a recent evolutionary history, but are not typical of our species. Being able to consume dairy products as an adult without getting sick, for instance, is a minority trait across all humanity, resulting from the coevolution of dairy farming and the LCT allele(s) that underlie adult lactose tolerance (Simoons 1970; Durham 1991). This character is a fundamental aspect of the biology of many humans, and one that has unquestionably evolved. We see no reason why the 'universal' evolved features of humanity merit more attention than the variable ones (Lewens 2012).

Nor do we find compelling the response that it is legitimate for one discipline (i.e. evolutionary psychology) to concentrate on studying typical aspects of humanity, since other fields (i.e. social anthropology) study human diversity (Machery 2008). Different academic fields utilize different research methods, hindering comparison. What is ultimately required is a comprehensive account of the human condition, providing a unified and coherent understanding of both common and diverse elements. There is a very real danger that this focus on human nature actively devalues an understanding of diversity—encouraging people to conclude that it captures the most important elements of the human condition. The evolutionary psychology community clearly believe themselves to be providing a comprehensive account of the key aspects of human psychology and behaviour, not just half of the story.

We are also concerned that the labelling of a subset of human-typical characters as 'human nature' would encourage researchers to jump to the erroneous conclusion that different processes explain what is relatively stable and what is variable (e.g. to attribute constancy to naturally selected genes and diversity to environmental factors). As we have seen, neither human development nor human evolution is as simple as that, and complex gene–environment interactions account for both the evolutionary history and the ontogeny of typical (e.g. reduced jaw muscle) and variable (i.e. adult lactose tolerance) characters.

In principle, the scientific community could agree a convention for how to deal with ambiguous or difficult traits, allowing a consensus to be reached on what is and what is not human nature. But to what end? The researchers concerned would have worked hard to select an arbitrary component of the human condition, capturing arbitrary elements of our species' evolutionary history and missing out on much of the interesting variation in human biology, behaviour, and cognition in the process, as well as much of the richness of the underlying causal dynamics—all to perpetuate an outdated conception of biological development.

## 7.3 Conclusions

Our title is a play on words. We suggest that 'human nature' is socially constructed in two senses. First, as social constructivists might complain, we believe that in the past scientists have devised diverse concepts of human nature to fulfil various roles, but that, on closer inspection, these bear little resemblance to biological reality and have little current scientific validity. The concept of human nature has been superseded and rendered obsolete by data that leaves it fundamentally unhelpful. Given the confusion surrounding the term, and its potential to seed further misinformation, like many others before us, we recommend that the concept of human nature be abandoned.

Second, as advocates of the niche-construction perspective in evolutionary biology (Odling-Smee et al. 2003), and recognizing the central role of cultural practices in allowing humans to modify and regulate their environments and those of their offspring (Stotz 2010; Kendal et al. 2011; Laland and O'Brien 2011; Sterelny 2012), we suggest that the human condition is the product of internal and external constructive processes operating over both developmental and evolutionary timescales. Many, perhaps most, aspects of human niche construction are the product of multiple individuals' activities—and hence are 'socially constructed'—and this self-built world has shaped both the evolution history of our species and the development of our children's behaviour and cognition. We recommend that evolutionarily minded researchers use these insights to endeavour to provide a comprehensive account of the human condition and of the multitude of processes through which it arose, focusing as much on diversity and on less prevalent features as on typicality.

## References

Badyaev, A. V., and Uller, T. (2009). 'Parental Effects in Ecology and Evolution: Mechanisms, Processes, and Implications.' *Philosophical Transactions of the Royal Society B* 364: 1169–77.

Bateson, P., and Mameli, M. (2007). 'The Innate and the Acquired: Useful Clusters or a Residual Distinction from Folk Biology?' *Developmental Psychobiology* 49: 818–31.

Bateson, P., and Martin, P. (1999). *Design for a Life: How Behaviour Develops*. London: Jonathan Cape.

Boakes, R. (1984). *From Darwin to Behaviourism: Psychology and the Minds of Animals*. Cambridge: Cambridge University Press.

Bolhuis, J., and Gahr, M. (2006). 'Neural Mechanisms of Birdsong Memory.' *Nature Reviews* 7: 347–57.

Boyd, R., and Richerson, P. J. (1985). *Culture and the Evolutionary Process*. Chicago: Chicago University Press.

Brown, D. E. (1991). *Human Universals*. New York: McGraw-Hill.

Buss, D. M. (1999). *Evolutionary Psychology: The New Science of the Mind*. Boston, Mass.: Allyn & Bacon.

Cashdan, E. (2013). 'What Is a Human Universal? Human Behavioral Ecology and Human Nature.' In S. M. Downes and E. Machery (eds), *Arguing about Human Nature: Contemporary Debates*, 71–80. London: Routledge.

Catchpole, C., and Slater, P. (2008). *Bird Song*. Cambridge: Cambridge University Press.

Cavalli-Sforza, L. L., and Feldman, M. W. (1981). *Cultural Transmission and Evolution: A Quantitative Approach*. Princeton, NJ: Princeton University Press.

Chiu, L., and Gilbert, S. F. (2015). 'The Birth of the Holobiont: Multi-Species Birthing through Mutual Scaffolding and Niche Construction.' *Biosemiotics* 8: 191–210.

Cochran, G., and Harpending, H. (2009). *The 10,000 Year Explosion*. New York: Basic Books.

Cryan, J. F., and Dinan, T. G. (2012). 'Mind-Altering Micro-Organisms: The Impact of the Gut Microbiota on Brain and Behaviour.' *Nature Reviews Neuroscience* 13: 701–12.

Danchin, E., Charmantier, A., Champagne, F. A., Mesoudi, A., Pujol, B., and Blanchet, S. (2011). 'Beyond DNA: Integrating Inclusive Inheritance into an Extended Theory of Evolution.' *Nature Reviews Genetics* 12: 475–86.

Daw, N. D., and Shohamy, D. (2008). 'The Cognitive Neuroscience of Motivation and Learning.' *Social Cognition* 26: 593–620.

Dean, L. G., Kendal, R. L., Schapiro, S. J., Thierry, B., and Laland, K. N. (2012). 'Identification of the Social and Cognitive Processes Underlying Human Cumulative Culture.' *Science* 335: 1114–18.

Durham, W. H. (1991). *Coevolution: Genes, Culture, and Human Diversity*. Stanford, Calif.: Stanford University Press.

Dusek, J. A., Out, H. H., Wohlhueter, A. L., Bhasin, M., Zerbini, L. F., Joseph, M. G., Benson, H., and Libermann, T. A. (2008). 'Genomic Counter-Stress Changes Induced by the Relaxation Response.' *PLoS ONE* 3: e2576.

Gangestad, S. W., and Simpson, J. A. (2000). 'The Evolution of Human Mating: Trade-Offs and Strategic Pluralism.' *Behavioral and Brain Sciences* 23: 573–644.

Gergely, G., and Csibra, G. (2005). 'The Social Construction of the Cultural Mind: Imitative Learning as a Mechanism of Human Pedagogy.' *Interaction Studies* 6: 463–81.

Gilbert, S. F. (2003). 'The Morphogenesis of Evolutionary Developmental Biology.' *International Journal of Developmental Biology* 47: 467–77.

Gilbert, S. F., and Epel, D. (2009). *Ecological Developmental Biology: Integrating Epigenetics, Medicine, and Evolution*. Sunderland, Mass.: Sinauer Associates.

Gottlieb, G. (1971). *Development of Species Identification in Birds*. Chicago: University of Chicago Press.

Green, R. E., Krause, J., Briggs, A. W., et al. (2010). 'A Draft Sequence of the Neanderthal Genome.' *Science* 328: 710.

Henrich, J., Heine, S. J., and Norenzayan, A. (2010). 'The Weirdest People in the World?' *Behavioral and Brain Sciences* 33: 61–3.

Herberstein, M. E. (2011). *Spider Behaviour: Flexibility and Versatility*. Cambridge: Cambridge University Press.

Hinde, R. A. (1982). *Ethology*. London: Fontana.

Hoppitt, W., and Laland, K. N. (2013). *Social Learning: An Introduction to Mechanisms, Methods, and Models*. Princeton, NJ: Princeton University Press.

Horn, G. (2004). 'Pathways of the Past: The Imprint of Memory.' *Nature Reviews Neuroscience* 5: 108–20.
Hull, D. L. (1986). 'On Human Nature.' *Proceedings of the Biennial Meeting of the Philosophy of Science Association* 2: 3–13.
Jablonka, E., and Lamb, M. (2014). *Evolution in Four Dimensions*. Cambridge, Mass.: MIT Press.
Jarvis, E. D., and Nottebohm, F. (1997). 'Motor-Driven Gene Expression.' *Proceedings of the National Academy of Sciences USA* 94: 3406–10.
Kaplan, H. S., and Gangestad, S. W. (2005). 'Life History Theory and Evolutionary Psychology.' In D. M. Buss (ed.), *The Handbook of Evolutionary Psychology*, 68–96. New York: Wiley.
Keller, E. F. (2010). *The Mirage of a Space between Nature and Nurture*. Durham, NC: Duke University Press.
Kendal, J., Tehrani, J. J., and Odling-Smee, F. J. (2011). 'Human Niche Construction in Interdisciplinary Focus.' *Philosophical Transactions of the Royal Society B* 366: 785–92.
Kirschner, M., and Gerhart, J. (2005). *The Plausibility of Life: Resolving Darwin's Dilemma*. New Haven, Conn.: Yale University Press.
Krause, J., Fu, Q., Good, J. M., et al. (2010). 'The Complete Mitochondrial DNA Genome of an Unknown Hominin from Southern Siberia.' *Nature* 464: 894–7.
Laland, K. N., and Brown, G. R. (2006). 'Niche Construction, Human Behaviour, and the Adaptive-Lag Hypothesis.' *Evolutionary Anthropology* 15: 95–104.
Laland, K. N., and Brown, G. R. (2011). *Sense and Nonsense: Evolutionary Perspectives on Human Behaviour*. Oxford: Oxford University Press.
Laland, K. N., and O'Brien, M. (2011). 'Cultural Niche Construction: An Introduction.' *Biological Theory* 6: 191–202.
Laland, K. N., Odling-Smee, J., Hoppitt, W., and Uller, T. (2012). 'More on How and Why: Cause and Effect in Biology Revisited.' *Biology and Philosophy* 28: 719–45.
Laland, K. N., Odling-Smee, J., Hoppitt, W., and Uller, T. (2013). 'More on How and Why: A Response to Commentaries.' *Biology and Philosophy* 28: 793–810.
Laland, K. N., Odling-Smee, F. J., and Myles, S. (2010). 'How Culture Has Shaped the Human Genome: Bringing Genetics and the Human Sciences Together.' *Nature Reviews Genetics* 11: 137–48.
Laland, K. N., Sterelny, K., Odling-Smee, F. J., Hoppitt, W., and Uller, T. (2011). 'Cause and Effect in Biology Revisited: Is Mayr's Proximate–Ultimate Dichotomy Still Useful?' *Science* 334: 1512–16.
Laland, K. N., Uller, T., Feldman, M. W., Sterelny, K., Müller, G. B., Moczek, A., Jablonka, E., and Odling-Smee, F. J. (2014). 'Does Evolutionary Theory Need a Rethink?' *Nature* 514: 161–4.
Laland, K. N., Uller, T., Feldman, M. W., et al. (2015). 'The Extended Evolutionary Synthesis: Its Structure, Assumptions, and Predictions.' *Proceedings of the Royal Society B* 282: rspb.royalsocietypublishing.org/content/282/1813/20151019.
Lehrman, D. S. (1953). 'Critique of Konrad Lorenz's Theory of Instinctive Behavior.' *Quarterly Review of Biology* 28: 337–63.
Lehrman, D. S. (1964). 'The Reproductive Behavior of Ring Doves.' *Scientific American* 211: 48–54.
Lewens, T. (2012). 'Human Nature: The Very Idea.' *Philosophy and Technology* 25: 459–74.
Lewens, T. (2015). *Cultural Evolution: Conceptual Challenges*, Oxford: Oxford University Press.

Lewontin, R. C. (1983). 'Gene, Organism, and Environment.' In D. S. Bendall (ed.), *Evolution from Molecules to Men*, 273–85. Cambridge: Cambridge University Press.
Lewontin, R. C. (2000). *The Triple Helix: Gene, Organism, and Environment*. Cambridge, Mass.: Harvard University Press.
Lorenz, K. (1965). *Evolution and Modification of Behavior*. Chicago: University of Chicago Press.
Machery, E. (2008). 'A Plea for Human Nature.' *Philosophy and Psychology* 21: 321–9.
Mameli, M., and Bateson, P. (2011). 'An Evaluation of the Concept of Innateness.' *Philosophical Transactions of the Royal Society B* 266: 436–43.
Mayr, E. (1961). 'Cause and Effect in Biology.' *Science* 134: 1501–6.
Nielsen, R., Hellmann, I., Hubisz, M., Bustamante, C., and Clark, A. G. (2007). 'Recent and Ongoing Selection in the Human Genome.' *Nature Reviews Genetics* 8: 857–68.
Nisbett, R. E. (2003). *The Geography of Thought*. New York: Free Press.
Noble, D. (2006). *The Music of Life*. Oxford: Oxford University Press.
Norenzayan, A., and Heine, S. J. (2005). 'Psychological Universals: What Are They and How Can We Know?' *Psychological Bulletin* 131: 763–84.
Odling-Smee, F. J., Laland, K. N., and Feldman, M. W. (1996). 'Niche Construction.' *American Naturalist* 147: 641–8.
Odling-Smee, F. J., Laland, K. N., and Feldman, M. W. (2003). *Niche Construction: The Neglected Process in Evolution*. Princeton, NJ: Princeton University Press.
Oyama, S. (1985). *The Ontogeny of Information*. Cambridge: Cambridge University Press.
Oyama, S., Gray, R., and Griffiths, P. (2001). *Cycles of Contingency: Developmental Systems and Evolution*. Cambridge, Mass.: MIT Press.
Perry, G. H., Dominy, N. J., Claw, K. G., et al. (2007). 'Diet and the Evolution of Human Amylase Gene Copy Number Variation.' *Nature Genetics* 39: 1256–60.
Pinker, S. (2003). *The Blank Slate: The Modern Denial of Human Nature*. New York: Penguin.
Richards, M. P., Schulting, R. J., and Hedges, R. E. M. (2003). 'Archaeology: Sharp Shift in Diet at Onset of Neolithic.' *Nature* 425: 366.
Richerson, P. J. (2014). 'Human Nature.' *EDGE*: www.edge.org/response-detail/25404
Richerson, P. J., and Boyd, R. (2005). *Not by Genes Alone: How Culture Transformed Human Evolution*. Chicago: University of Chicago Press.
Richerson, P. J., Boyd, R., and Henrich, J. (2010). 'Gene–Culture Coevolution in the Age of Genomics.' *PNAS* 107: 8985–92.
Ridley, M. (2003). *Nature via Nurture: Genes, Experience, and What Makes Us Human*. New York: HarperCollins.
Roca, A. L., Georgiadis, N., Pecon-Slattery, J., and O'Brien, S. J. (2001). 'Genetic Evidence for Two Species of Elephant in Africa.' *Science* 293: 1473–7.
Sabeti, P. C., Varilly, P., Fry, B., et al. (2007). 'Genome-Wide Detection and Characterization of Positive Selection in Human Populations.' *Nature* 449: 913–18.
Scott-Phillips, T. C., Laland, K. N., Shuker, D. M., Dickins, T. E., and West, S. A. (2014). 'The Niche Construction Perspective: A Critical Appraisal.' *Evolution* 68: 1231–43.
Segerstråle, U. (2000). *Defenders of the Truth: The Sociobiology Debate*. Oxford: Oxford University Press.
Simoons, F. (1970). 'Primary Adult Lactose Intolerance and the Milking Habit: A Problem in Biologic and Cultural Interrelations II: A Culture Historical Hypothesis.' *American Journal of Digestive Diseases* 15: 695–710.

Slater, P. J. B. (1978). *Sex Hormones and Behaviour*. Baltimore, Md.: University Park Press.

Stedman, H. H., Kozyak, B. W., Nelson, A., et al. (2004). 'Myosin Gene Mutation Correlates with Anatomical Changes in the Human Lineage.' *Nature* 428: 415–18.

Sterelny, K. (2012). *The Evolved Apprentice*. Cambridge, Mass.: MIT Press.

Stotz, K. (2010). 'Human Nature and Cognitive-Developmental Niche Construction.' *Phenomenological and Cognitive Science* 9: 483–501.

Sultan, S. E. (2015). *Organism and Environment*. Oxford: Oxford University Press.

Tennie, C., Call, J., and Tomasello, M. (2009). 'Ratcheting up the Ratchet: On the Evolution of Cumulative Culture.' *Philosophical Transactions of the Royal Society B* 364: 2405–15.

Tinbergen, N. (1963). 'On Aims and Methods of Ethology.' *Zeitschrift für Tierpsychologie* 20: 410–33.

Tomasello, M. (1999). *The Cultural Origins of Human Cognition*. Cambridge, Mass.: Harvard University Press.

Tooby, J., and Cosmides, L. (1992). 'The Psychological Foundations of Culture.' In J. H. Barkow, L. Cosmides, and J. Tooby (eds), *The Adapted Mind*, 531–49. New York: Oxford University Press.

Tooby, J., and Cosmides, L. (2005). 'Conceptual Foundations of Evolutionary Psychology.' In D. Buss (ed.), *The Handbook of Evolutionary Psychology*, 5–67. Hoboken, NJ: Wiley.

Voight, B. F., Kudaravalli, S., Wen, X., and Pritchard, J. K. (2006). 'A Map of Recent Positive Selection in the Human Genome.' *PLoS Biology* 4: e72.

Wang, E. T., Kodama, G., Baldi, P., and Moyzis, R K. (2006). 'Global Landscape of Recent Inferred Darwinian Selection for *Homo sapiens*.' *Proceedings of the National Academy of Sciences USA* 103: 135–40.

West-Eberhard, M. J. (2003). *Developmental Plasticity and Evolution*. Oxford: Oxford University Press.

Wilson, E. O. (1994). *Naturalist*. Washington, DC: Island Press.

Wrangham, R. W., Jones, J. H., Laden, G., Pilbeam, D., and Conklin-Brittain, N. (1999). 'The Raw and the Stolen.' *Current Anthropology* 40: 567–94.

Zeigler, H. P., and Marler, P. (2008). *Neuroscience of Birdsong*. Cambridge, Mass.: MIT Press.

# 8

# The Use and Non-Use of the Human Nature Concept by Evolutionary Biologists

*Peter J. Richerson*

*What does evolution teach us about human nature? It tells us that human nature is a superstition.*

Ghiselin (1997: 1)

## 8.1 Introduction

'Human nature' seems on the face of it to be an essentialist concept with no place in a proper Darwinian analysis of human behaviour. David Hull (1986) wrote a powerful critique along these lines. Tim Lewens (2012) endorses and extends Hull's analysis, but points out that a very weak version of human nature is tenable, one which just refers to the ensemble of genetic differences between ourselves and other species. Lewens argues that any strong conception of human nature in which the term does real work—his examples are evolutionary psychology and the morality of altering our supposed nature—cannot be sustained. Nevertheless, the term 'human nature' figures heavily in the writings of many of the most important scholars over the last forty years who want to use evolutionary theory to explain human behaviour.

The aim of this chapter is to examine a sample of the most prominent papers and books using the term 'human nature' in what seems to be a strong way, and contrast them with human evolutionists who do not use the term at all, or at least not in a strong way. Several of the human nature pieces I will examine were written by eminent evolutionary biologists who have taken an interest in human evolution. The others are by evolutionary psychologists whose knowledge of evolution is better than adequate. They would not defend a simplistic typological notion of human nature, though some arguments could almost be construed as such. They use the term in titles of books and papers, suggesting that it does do real work in their thinking. What work does it do? Does it do this work successfully? How do the recent users of human nature fit into the larger picture of evolutionarily literate accounts of human behaviour? Do users and non-users of the term have significantly different ideas about the processes active in human evolution? I will proceed in an historical fashion, starting with Darwin, whose ideas about human evolution in the *Descent of Man* were rather detailed and sophisticated.

The main issue turns out to be how strongly authors are committed to the tenets of the Modern Synthesis. The authors who use the term 'human nature' in its strong form are committed to natural selection acting on genes being the 'ultimate' explanation for human evolution. Ernst Mayr's famous paper 'Cause and Effect in Biology' (1961) argued that causes in biology could be divided into ultimate causes, like those resulting from natural selection on genes, and proximate causes, those resulting from physiology and mechanisms of phenotypic flexibility like individual learning. He held that ultimate causes explain how proximate causes evolve, but that proximate causes play no role in explaining how ultimate causes like selection operate. If human culture is just a mechanism of phenotypic flexibility, then it cannot be a part of the ultimate explanation of human behaviour. The human nature theorists reviewed here want to collect all the results of selection acting on human genes under the term 'human nature', providing the ultimate answer to why humans are the creatures they are.

Those of us who use 'human nature' in Lewens's weak sense, if at all, doubt that non-genetic inheritance systems like human culture can be treated as proximate causes that have no effect on human genetic evolution (Mesoudi et al. 2013; Richerson and Boyd 2005). First, human culture can be subject to natural selection just as genes can, so culture is part of the ultimate causal process. Second, we have examples of culture acting as a selective force on genes (Laland et al. 2010; Ross and Richerson 2014). More broadly, culture may well have shaped many innate aspects of human psychology, physiology, and anatomy by culture-led gene–culture coevolution (Richerson et al. 2016). Cultural evolution is affected by forces considered unimportant in the Modern Synthesis, such as the Lamarckian inheritance of acquired variation. If cultures can act as selective forces on genes, then the toolkit of the Modern Synthesis is not adequate to the task of providing a complete account for human evolution. The broader version of this critique, based on other non-genetic inheritance systems and applying to many species, is called the extended evolutionary synthesis (Laland et al. 2011).[1]

I will argue that the Modern Synthesis human nature theorists are on the horns of a dilemma. They, for the most part, do not deny that human culture or human reason have played a role in our evolutionary history; but, they maintain, it can't be a truly fundamental role because cultural evolution plainly has features that are outside the scope of the Modern Synthesis, such as the inheritance of acquired variation. The commonest strategy to try to avoid the dilemma is to imagine that natural selection dominated our evolution until quite recently and that it is only today, or fairly recently, that we have enough knowledge to affect our own genetic evolution. Culture is like the paint on a house, the last and least important feature of its construction, however

---

[1] For models that illustrate the basic dynamic differences between the Modern Synthesis and the gene–culture coevolutionary treatments of culture, see Morgan (2016).

pleasing to the eye. The trouble with this solution is that the expanding importance of culture and innovation is the main story of the evolution of our genus *Homo* going back more than two million years. Arguably, culture driven gene–culture coevolution was foundational in human evolution. The solutions to the dilemma proposed by human nature theorists do not seem plausible to me.

## 8.2 Darwin

It is interesting to consider contemporary writings on human nature in light of Darwin's ideas. In *The Descent of Man, and Selection in Relation to Sex* (1874), Darwin laid out his ideas about human evolution. In the preface to the second edition, he writes that 'great weight must be attributed to the inherited effects of use and disuse, with respect both to the body and the mind'. He protests that he entertains the importance of many evolutionary processes besides random variation and natural selection.

Two important themes are relevant to twentieth- and twenty-first-century writing on human nature. First, in chapter 5, entitled 'The Improvement of the Intellectual and Moral Faculties during Primeval and Civilized Times', he distinguishes between the processes important in primeval and civilized times; during primeval times, natural selection would have been the primary force advancing the intellectual and moral faculties. He also remarks that 'as soon as the progenitors of man became social (and this probably occurred at a very early period) the principle of imitation, and reason, and experience would have increased, and much modified the intellectual powers in a way of which we see only traces in lower animals' (1874: 174). Even during the primeval period, he imagines that what we would summarize today as intellectual and cultural factors were already important:

> I have already said enough, while treating of the lower races, on the causes which lead to the advance of morality, namely, the approbations of our fellow men—the strengthening of our sympathies by habit—example and imitation—reason—experience, and even self-interest—instruction during youth, and religious feelings. (Darwin 1874: 185–6)

As regards 'civilized times', the role of what we now call cultural factors was even more important:

> With highly civilized nations, continued progress depends in a subordinate degree on natural selection [...] The more efficient causes of progress seem to consist of a good education during youth while the brain is impressible, and of a high standard of excellence, inculcated by the ablest and best men, embodied in the laws, customs, and traditions of the nation, and enforced by public opinion. (Darwin 1874: 192)

Second, in his chapter 'On the Races of Man', Darwin argued in favour of the psychic unity of humans and against the idea that the races could be considered separate species (a major plank in the racist ideology of his day):

> Although the existing races differ in many respects, as in color, hair, shape of the skull, proportions of the body, etc., yet, if their whole structure be taken into consideration, they are found

to resemble each other closely on a multitude of points. Many of these are so unimportant or of so singular a nature that it is extremely improbable that they should have been independently acquired by aboriginally distinct species or races. The same remark holds good with equal or greater force with respect to the numerous points of mental similarity between the most distinct races of man. The American aborigines, Negroes, and Europeans are as different from each other in mind as any three races that can be named; yet I was constantly struck, while living with the Fuegans on board the *Beagle*, with the many little traits of character showing how similar their minds were to ours; and so it was with a full-blooded Negro with whom I happened once to be intimate. (Darwin 1874: 231–40)

Darwin does use terms like 'higher and lower races' and similar illiberal characterizations of non-Europeans that seem in glaring contradiction to this quote. Gruber and Barrett (1974) note these discrepancies in Darwin's writing on humans. They analyse his notebooks and private letters, and argue that he retained throughout his life a partly justified fear of persecution for his materialistic views. He larded his public prose with conventional Victorian tropes as a kind of protective colouration for his more radical opinions, and was privately upset with himself for having done so in one letter they quote.

The term 'human nature' is absent from the detailed index of *The Descent of Man*. Clearly, Darwin saw inherited variation everywhere; and if he had used the phrase, it would likely have been in Lewens's permissible but powerless vein. Darwin's own ideas on human evolution were influential in the late nineteenth century, but had little influence on the emerging social science disciplines as they crystallized at the turn of the twentieth (Hodgson 2004; Richards 1987). Nor do most evolutionary biologists or Evolutionary Psychologists writing about human evolution cite *The Descent of Man* in any detail.[2] From the viewpoint of evolutionists writing after the mid-twentieth-century Modern Synthesis (see Huxley 1942), Darwin's subscribing to the inheritance of acquired variation was perhaps his greatest error. Thus evolutionary biologists came to the problem of human evolution with a worldview strongly influenced by genes as the dominant, if not the exclusive, means of inheritance. Social scientists were aware that culture could be viewed as a Lamarckian sort of inheritance system (Kroeber 1948), but until the late twentieth century they did not develop this idea with anything like the theoretical and empirical intensity with which evolutionary biologists had pursued the gene-based view of organic evolution.

## 8.3 Theodosius Dobzhansky

Dobzhansky was the architect of the Modern Synthesis who took the most interest in human evolution. His prize-winning book *Mankind Evolving: The Evolution of the Human Species* (1962) had gone through seventeen printings by 1975, when it suddenly

---

[2] The capitalized 'Evolutionary Psychologists' refers to the influential school of evolutionary psychology established by John Tooby and Leda Cosmides. Steven Pinker is a well-known adherent. But many evolution-minded psychologists are not Evolutionary Psychologists in this sense.

fell into the shadow of important developments in the evolution of social behaviour that E. O. Wilson summarized in *Sociobiology: The New Synthesis* (1975). Dobzhansky used the term 'human nature' in Lewens's innocent sense, but it does not appear in the index. *Mankind Evolving* is notable for its attempt to reconcile the then dominant view of genetic inheritance as fundamental to the organic evolution of all species including humans with the then well-established view of most social scientists that the dominant source of variation in humans was cultural. He introduces his view thusly:

The thesis to be set forth in the present book is that man has both a nature and a 'history'. Human evolution has two components, the biological or organic, and the cultural or superorganismic. These components are neither mutually exclusive nor independent but interrelated and interdependent. Human evolution cannot be understood as a purely biological process, nor can it be adequately described as a history of culture. It is the interaction of biology and culture. There exists a feedback between biological and cultural processes.   (Dobzhansky 1962: 18)

He goes on to critique biologists who disparage the importance of culture and social scientists who want biological evolution to have merely evolved a vehicle for cultural evolution. He introduces some science fiction examples of how cultural evolution of the science of human biology might eventually produce human populations without the need for sexual reproduction; culture could be that powerful.

Other passages in his introduction reinforce these thoughts:

In producing the genetic basis for culture, biological evolution has transcended itself—it has produced the superorganic. Yet the superorganic has not annulled the organic.   (1962: 20)

The fact which must be stressed, because it has been frequently missed or misrepresented, is that biological and cultural evolution are parts of the same natural process.   (1962: 22)

Dobzhansky's thesis sounds a lot like Darwin in his *Descent* (1874) with the now strongly established genetic inheritance theory substituted for Darwin's ideas of inheritance, which were, by his own admission, unsatisfactory speculations. Interestingly, *The Descent of Man* is not formally cited in *Mankind Evolving*, although at one point Dobzhansky remarks (1962: 7): 'In his books Darwin confined himself to biological matters, even in the *Descent of Man* and *The Expression of Emotion in Man and Animals*' (1962: 7). As we have seen, this is quite incorrect, suggesting that Dobzhansky had never read *The Descent* carefully or, if he did, politely passed over those of Darwin's ideas that were incompatible with the Modern Synthesis.

*Mankind Evolving* is full of interesting discussions on a great variety of topics. Its treatment of human organic evolution and genetics is detailed and authoritative for its day, as one might expect from a great evolutionist. It engages well with the social sciences, for example, discussing Freudian psychotherapy, the role of genes and culture in race, caste in India, the nature/nurture controversy, the feedback between genes and culture, among many other live issues of the time. He sympathetically cites social scientists, historians, and humanistic writers. He does insist that evolution of genes was and continues to be important, even if genetic evolution has transcended itself to produce the superorganic.

The difficulty is that the discussions of cultural evolution never climb down from the lofty generalities and vaguely defined terms evident in the passages cited above. Just how did the transcendence happen? What exactly is involved in the superorganic? What is the natural process of which both biological and cultural evolution are a part? Some of the problem was the fact, noted above, that social scientists after 1900 had not followed up on Darwin's programme for the study of culture as a Darwinian evolutionary system. There was simply no large body of literature like that in evolutionary genetics with which to build its cultural partner. Furthermore, Dobzhansky's political beliefs and personal style probably led him to have a distaste for challenging social scientists on such issues. He co-authored a paper in *Science* with Ashley Montague (Dobzhansky and Montagu 1947), an anti-racist activist with whom he worked on the UNESCO statement *The Race Question* following the Second World War. Montagu was not inclined to view genetic differences between groups as important at all, but Dobzhansky and other biologists reviewing the draft statement were more cautious (Brattain 2007). They could see that some genetic differences between human groups were real, although few of them were well understood at the time. Although Dobzhansky and the other UNESCO biologists wanted to forestall invidious distinctions based on biological differences, they also wanted to make sure that human biologists were not to be condemned as racists just because they were investigating biological differences between human groups. I interpret the result as a sort of political compromise between biologists and social scientists. As Dobzhansky and Montagu (1947: 590) put it:

The effect of natural selection in man has probably been to render genotypic differences in personality traits, as between individuals and particularly as between races, relatively unimportant compared to phenotypic plasticity. Instead of having his responses genetically fixed as in other animal species, man is a species that invents its own responses, and it is out of this unique ability to invent, to improvise, his responses that his culture is born.

This formulation gives human biologists plenty of room to work while granting that the study of human culture by social scientists and humanists was also an entirely legitimate enterprise. Everyone could do their customary work without stepping on toes. When the next generation of scholars tried to build concrete models of transcendence, the superorganic, and a common natural basis of cultural and genetic evolution, the Dobzhansky–Montagu compromise immediately led to serious intellectual conflicts, some of them highly politicized, for example Jensen's (1969) argument that racial differences in IQ in the US are largely genetic.

The immediate cause of the rise in the use of the term 'human nature' was the seminal advances in the theory of behavioural evolution, including W. D. Hamilton's (1964) model of inclusive fitness, George Williams (1966) casting doubt on group selection as an explanation of helping behaviour, Robert Trivers' (1971) paper on reciprocal altruism, Richard Alexander's (1974) synthetic review, and John Maynard Smith's (1976) application of game theory to evolutionary questions. Now evolutionists had real tools to apply to the study of behaviour. The last chapter of Edward O. Wilson's magisterial *Sociobiology* (1975) showed how these tools could be applied to explaining human behaviour from an evolutionary perspective. The negative reaction to this on the part of certain left-wing biologists and social scientists is a famous episode that has coloured relations between the natural and social sciences to this day (Segerstrale 2000). Dobzhansky and Montague's peace treaty, if that is the right term, broke down almost completely.

## 8.4 Edward O. Wilson

Wilson's *On Human Nature* (1978) enlarges upon his last chapter of *Sociobiology*. No biologist as sophisticated as Wilson would have an essentialist view of humans; but by using the term as his title, he must have thought it could do work for him. In the preface to the reprint of the book (2004: ix–x), he summarizes his 1978 position:

[The naturalistic view] held that the brain and mind are entirely biological in origin and have been highly structured through evolution by natural selection. Human nature exists, composed of the complex biases of passion and learning propensities often loosely referred to as instincts. The instincts were created over millions of years, when human beings were Paleolithic hunter-gatherers. As a consequence, they still bear the archaic imprint of our species' biological heritage. Human nature can thus be ultimately understood only with the aid of the scientific method. Culture evolves in response to environmental and historical contingencies, as common sense suggests, but its trajectories are powerfully guided by the inborn biases of human nature.

In his first chapter, Wilson emphasizes that the brain is a purely biological implement that includes direct adaptations to past environments and innate censors and motivators that control culture. He speaks of reductionism, yet also of emergent phenomena; human behaviour must conform to the laws of biology, but also might transcend them. This chapter differs little from Dobzhansky's (1962) position, except for his invocation of the 'human nature' term. In the rest of the book he explains that he means something rather stronger than Dobzhansky's position.

The second chapter contains a discussion of both human universals and genetic variation in important traits, much as Dobzhansky (1962) did. Wilson is sympathetic to the innatist cognitive science picture of the mind, and argues against the *tabula rasa* view. He doubts the behaviourist approach to understanding humans that imagines them to have been substantially shaped by general-purpose learning mechanisms. It is a little difficult to know how to take these passages arguing against the blank slate. It is quite hard for anyone with any commitment to evolutionary biology to believe that genes play no role in the properties of the mind/brain. Behaviourists, often falsely associated with the idea of a blank slate, who want to talk about the environmental control of behaviour know well that the behaviour of different species and even different individuals is reinforced by different stimuli; the wolf and the moose may live in the same environment, but the wolf's behaviour is not reinforced by browsing on plants and the moose's behaviour is not reinforced by eating wolves. Furthermore, behaviourists are perfectly comfortable with innate reflexes and fixed action patterns (Baum 2005: ch. 4). Rejecting the *tabula rasa* is to reject a near-straw man argument, held in pure form by perhaps only a few social scientists ignorant of modern biology. But rejecting a straw man argument is no support at all for a strongly innatist one. Cultural evolutionists, like behaviourists, are quite well aware that genes and genetic evolution have a large role to play in explaining human behaviour. They simply hold that cultural evolution and a history of reinforcement are important as well.

A remark Wilson makes is interesting in the light of later developments I discuss in the penultimate section of this chapter:

> Let me now rephrase the central proposition in a somewhat stronger and more interesting form: if the genetic components of human nature did not originate by natural selection, fundamental evolutionary theory is in trouble. At the very least the theory of evolution would have to be altered to account for new and as yet unimagined form of genetic change in populations. Consequently, an auxiliary goal of human sociobiology is to learn whether the evolution of human nature conforms to conventional evolutionary theory. The possibility that the effort will fail conveys to more adventurous biologists the not unpleasant whiff of grapeshot, a crackle of thin ice. (Wilson 1978: 33–4)

It is not clear if Wilson considers himself to be an adventurous biologist in this context, but he is fairly clear that if anyone imagines that culture could somehow play an important selective role in human genetic evolution, it would represent a revolutionary challenge to the Modern Synthesis. If culture evolves in a partly Lamarckian way, or if culture exerts selection on genes via social selection, or if natural selection operates via culturally constructed environments, then the Modern Synthesis toolkit is incomplete, at least in the human case. Evolutionary biologists had struggled in the 1930s to disprove the idea of the inheritance of acquired variation and correct Darwin's 'greatest error'. They did not follow Darwin's lead in giving cultural processes a fundamental role in human evolution. In so doing, they 'hardened the Synthesis' in the eyes of some (Burian 1988; Gould 1982). What the cultural evolutionists threaten, it seems, is not merely a minor amendment to the Modern Synthesis, appropriate for one peculiar species (which is how I viewed it at the time). It threatens a major tenet of the Modern Synthesis.

Wilson invokes the theme that we already saw in Dobzhansky, a temporal subdivision of human evolution into a long period when natural selection was dominant and a short, recent period when culture became important:

> We can be fairly certain that most of the genetic evolution of human social behavior occurred over the five million years prior to civilization, when the species consisted of sparse, relatively immobile populations of hunter-gatherers. On the other hand, by far the greater part of cultural evolution has occurred since the origin of agriculture and cities approximately 10,000 years ago. Although genetic evolution of some kind continued during this latter, historical sprint, it cannot have fashioned more than a tiny fraction of the traits of human nature. (Wilson 1978: 34)

Contrast this with Darwin's picture that imitation, reason, and experience began to modify the human intellect early in human evolution.

Later, Wilson articulates a strongly innatist picture of development:

The newborn [infant's brain] is now seen to be wired with awesome precision. [...] This marvelous robot [the infant] is launched into the world under the care of its parents. Its rapidly accumulating experience will soon transform it into an independently thinking and feeling individual. Then the essential components of social behavior will be added—language, pair bonding, rage at ego injury, love, tribalism, and all the remainder of the human-specific repertory. But to what extent does the wiring of the neurons, so undeniably encoded in the genes, preordain the directions that social development will follow?   (Wilson 1978: 54–5)

Wilson has a complex answer to this question. Human behaviour is certainly not composed entirely of reflexes and fixed action patterns the way insects' behaviour seems to be: 'Human genes prescribe the capacity to develop a certain array of traits' (1978: 56). The degree and kind of prescription is variable depending upon the trait, and pathological conditions that affect prescriptions frequently have a strong heritable component, such as schizophrenia. Some traits, such as the expression of emotions, are practically invariant. He rejects the alleged behaviourist idea that human learning gives rise to unlimited flexibility in response to environmental contingencies: 'The learning potential of each species appears to be fully programmed by the structure of the brain, the sequence of release of its hormones, and, ultimately, its genes' (p. 65). He comments favourably on the theory of development put forward by Jean Piaget, the leading child development researcher of the time:

Piaget, who was originally trained as a biologist, views intellectual development as the interaction of an inherited genetic program with the environment. It is no coincidence that he calls this conception 'genetic epistemology', in effect the study of the hereditary unfolding of understanding.   (Wilson 1978: 66–7)

Wilson offers many examples of human universals in support of these ideas. Recent work on brains tends to emphasize the flexibility and responsiveness of brain development to environmental factors (e.g. Krubitzer and Seelke 2012). Recent work on child development tends to emphasize that children are adept social learners and that development is highly structured by their cultures (e.g. Carey 2009).

In chapter 4, Wilson gives culture what he thinks is its due:

When societies are viewed strictly as populations, the relationship between culture and heredity can be defined more precisely. Human evolution proceeds along a dual track of inheritance: cultural and biological. Cultural evolution is Lamarckian and very fast, whereas biological evolution is Darwinian and usually very slow.   (Wilson 1978: 78)

By 'Darwinian' here, Wilson means it in the Modern Synthesis, rather than in Darwin's, sense. He remarks that 'Lamarckism has been entirely discounted as the basis of biological evolution, but it is precisely what happens in the case of cultural evolution'

(p. 79), showing again his conventional commitment to the Modern Synthesis. A key passage follows (p. 79):

> Because it is far slower than Lamarckian evolution, biological evolution is always quickly outrun by cultural change. Yet the divergence cannot become too great, because ultimately the social environment created by cultural evolution will be tracked by biological natural selection.

This theme is much elaborated by Lumsden and Wilson in their *Genes, Mind, and Culture: The Coevolutionary Process* (1981: ch. 6). They develop a mathematical model of what they consider to be the 'complete coevolutionary circuit', which they try to capture by their famous leash metaphor of the genetic master controlling the cultural dog.

In some passages, Wilson does seem to invoke cultural evolution as a creative process in human evolution:

> the earliest men or man-apes started to walk erect when they came to spend most or all of their time on the ground. Their hands were freed, the manufacture and handling of artifacts was made easier, and intelligence grew as the tool-using habit improved. With mental capacity and the tendency to use artifacts increasing through mutual reinforcement, the entire materials-based culture expanded. Now the species moved onto the dual track of evolution: genetic evolution by natural selection enlarged the capacity for culture and culture enhanced the genetic fitness of those who made maximum use of it. Cooperation during hunting was perfected and provided a new impetus for the evolution of intelligence, and so on through repeated cycles of causation. (Wilson 1978: 85)

Here, cultural evolution is doing all the work that contemporary gene–culture coevolutionists would attribute to it; but they would additionally argue that the Lamarckian properties of culture and other processes outside the Modern Synthesis have to be introduced into models of the evolutionary processes. To be plausible, the human nature account needs cultural evolution to be as Darwin imagined it, but the human nature view derived from the Modern Synthesis forbids that. If the genetic leash picture is correct, culture is under the same ultimate genetic control as any other mechanism of phenotypic flexibility, and the Modern Synthesis is safe. Another version of the leash metaphor, hinted at by Wilson's observation that cultural evolution is faster than genetic evolution, is that the cultural dog is not only fast but strong. The cultural dog might be walking human nature as much as vice versa; American readers will be familiar with the comic *Marmaduke* in which a Great Dane drags his human walkers into humorous situations by their mutual leash.

I believe that Wilson's human nature position is on the horns of a dilemma. If culture lacks the key evolutionary properties of the Modern Synthesis account of evolution, then it can't do what the above passage requires; but if it does have these properties, then how can natural selection on genes be guaranteed to track cultural evolution and keep it tame at the end of a genetic leash? Won't fast cultural evolution by social selection and niche construction impose on genetic evolution? The question then is whether any of the later authors who have espoused the human nature concept are free of this dilemma.

Most of the rest of *On Human Nature* takes up major thematic issues—aggression, sex, altruism, and religion—pointing out that these categories of behaviour have roots in pre-human behaviour and quite plausibly have important genetic influences. They also invoke cultural factors, such as group selection at the level of whole cultural groups and rational thought in the chapter on aggression:

> Although the evidence suggests that the biological nature of humankind launched the evolution of organized aggression and roughly directed its early history across many societies, the eventual outcome of evolution will be determined by cultural processes brought under the control of rational thought. (Wilson 1978: 116)

This is not too different from Darwin's (1874: 185–6) proposal that factors like reason played a role in moral progress. Cultural evolutionists like myself also see reason playing a role in human behaviour: humans often carefully consider which cultural variants to adopt, and sometimes their thinking leads them to create new variations by modifying an existing variant rather than developing a new one. Trouble arises with Wilson's human nature idea when we consider the question of when in human evolution reason began to play a role in directing human behaviour. If the influence of reason comes early enough, it introduces the process of culture-driven gene–culture coevolution deep into human history. Given that our great ape ancestors were already rather big-brained animals and that progressively increasing brain size characterizes our genus throughout the Pleistocene, much as the Wilson (1978: 85) quote above suggests, reason and culture could quite plausibly have generated social selection and niche construction effects on human genes for much, or potentially all, of the history of *Homo*. Humans have artificially selected the genes of domesticates following purposive procedures for ten millennia. Half-reasonable and moderately cultural early *Homo* might have used purposive procedures to select mates and hunting partners, and to sanction enemies. Such behaviour would act as a selective force on any genetic variation for cooperative abilities. Humans in effect might have domesticated themselves via social selection (Boehm 2008).

Social selection is the effect on the genetic fitness of actors as a result of the rewards or punishments that other actors impose on them. Social selection might be a mechanism by which culturally transmitted norms strongly select for genes, leading to the easier conformity of behaviour to norms. Arguably, human populations have relatively low frequencies of psychopathy, because we have been living for much of our evolution in norm-bound groups that have rewarded cooperation and punished those who could not follow norms (Chudek and Henrich 2011).

Wilson is comfortable with group selection, perhaps even culture-based group selection, playing a role in the evolution of human aggression:

> The evolution of warfare was an autocatalytic reaction that could not be halted by any people, because to attempt to reverse the process unilaterally was to fall victim. A new mode of natural selection was operating at the level of entire societies [...] as societies become more centralized and complex, they develop more sophisticated military organizations and techniques of battle,

and the greater their military sophistication, the more likely they are to expand their territories and to displace competing cultures. (Wilson 1978: 116–17)

But there is no reason to think that this process is restricted to relatively modern societies. Darwin (1874: 175) supposed that conflict between ancient tribes would have favoured courage, sympathy, faithfulness, and mutual aid, because tribes more endowed with such virtues would out-compete those less endowed. We know from patterns of languages, artefacts, and violence in the archaeological record that human groups moved, expanded, and contracted long before human societies became centralized and complex (e.g. Jorgensen 1980; Schwitalla et al. 2014; Tostevin 2013).

The final chapter of *On Human Nature* is entitled 'Hope'. A human nature evolved for hunting and gathering in the Pleistocene might not seem likely to offer much ground for hope. We may try earnestly to promote peace via reason and culture, but isn't our behaviour ultimately governed by a human nature that is innately aggressive? Wilson states the issue as a circularity with a solution:

we are forced to choose among the elements of human nature by reference to the value systems which these same elements created in an evolutionary age now long vanished. Fortunately, this circularity of the human predicament is not so tight that it cannot be broken through an exercise of will. The principle task of human biology is to identify and to measure the constraints that influence the decisions of ethical philosophers and everyone else, and to infer their significance through neurophysiological and phylogenetic reconstructions of the mind. This enterprise is a necessary complement to the continued study of cultural evolution. It will alter the foundation of the social sciences but in no way diminish their richness and importance. In the process it will fashion a biology of ethics, which will make possible the selection of a more deeply understood and enduring code of moral values. (Wilson 1978: 196)

Once again, if an exercise of will can break the circularity nowadays, then why should this not also be true for an evolutionarily relevant span of the past? The famously fierce Comanche were also adept peace-makers and alliance-builders when it suited their interests (Hämäläinen 2008).

To summarize, *On Human Nature* defends a concept of human nature that boils down to these four claims:

1. Natural selection shaped human genes for most of human evolution with little influence from cultural processes.
2. Naturally selected genes powerfully influence human development, irrespective of cultural differences.
3. Cultural evolution, and the effects of such forces as reason, are a recent development in human evolution.
4. The scientific study of human nature will permit us to work around its limitations.

Wilson's concept of human nature is not essentialist; it is entirely based on Modern Synthesis evolutionary biology as applied to social behaviour, so it cannot be objected to on the grounds that Hull (1986) articulated. Indeed, it privileges Modern Synthesis evolutionary processes to the point of denying any fundamental contribution of cultural evolutionary processes to the evolution of human nature. In this sense, the outline of the argument is very similar to Dobzhansky's, although less tolerant of the social sciences' general disinterest in genes and evolution. Like Dobzhansky, it does give great weight to cultural processes in recent periods, under the control of reason, and especially when guided by science. In his tools-and-brains scenario, Wilson does introduce an element of gene–culture coevolution, apparently not thinking this is a problem for a Modern Synthesis account. These concessions to cultural processes are necessary to explain the emergence of complex cultural adaptations and to offer a path to improving the human condition. In Lumsden and Wilson's (1981) 'complete coevolutionary circuit', genetic evolution can adapt to cultural changes and force them to be consistent with selection acting on genes, but cultural processes following their imperatives cannot impose themselves on genes. The trouble is that this picture doesn't offer a principled reason to restrict the operation of cultural processes to a very late phase of human evolution when it is fairly clear that culture-driven gene–culture coevolutionary processes operated fairly deep into the history of *Homo*, if not throughout the whole history of the genus, as cultural evolutionists are prone to think (Ross and Richerson 2014). Indeed, Wilson's scenario of tool evolution driving the growth of the intellect does imagine that a form of cultural evolution drove biological evolution. If so, then the Modern Synthesis, as extended to behavioural phenomena by Hamilton, Trivers, and Maynard Smith, did not furnish all the evolutionary tools needed to explain the evolution of the human genome any more than the human genome gives us all the tools to understand culture (Morgan 2016).

## 8.5 Richard D. Alexander

Alexander (1974; see also his 1979, 1987) wrote a strong polemic for the then-recent advances in the evolutionary theory of social biology:

For several years the study of social behavior has been undergoing a revolution with far-reaching consequences for the social and biological sciences. Partly responsible are three recent changes in the attitudes of evolutionary biologists. First was growing acceptance of the evidence that the potency of natural selection is overwhelmingly concentrated at levels no higher than that of the individual. Second was revival of the comparative method, especially as applied to behavior and life histories. Third was spread of the realization that not only are all aspects of structure and function of organisms to be understood solely as products of selection, but because of their peculiarly direct relationship to the forces of selection, behavior and life history phenomena, long neglected by the evolutionists, may be among the most predictable of all phenotypic attributes.   (Alexander 1974: 325)

Mention of the human case in his broad comparative discussion occur frequently, under such headings as individual versus group selection, kin selection and nepotism, the evolutionary forces favouring group living, parental investment and manipulation of progeny, and social interactions in different kinds of social groups. Since the theory for interpreting all the cases is based on the work of Hamilton, Trivers, and like-minded authors, the adherence to the amended Modern Synthesis orthodoxy is conspicuous. So it is rather remarkable that, in the end, Alexander entertains the possibility of group selection in the human case:

> I began with a denial of any great significance for the phenomenon of group selection. It is appropriate, perhaps, to finish with a caveat. For two reasons human social groups represent an almost ideal model for potent selection at the group level. First, the human species is (and possibly always has been) composed of competing and essentially hostile groups that frequently have not only behaved toward one another in the manner of different species, but also have been able quickly to develop enormous differences in reproductive and competitive ability because of cultural innovation and its cumulative effects. Second, human groups are uniquely able to plan and act as units, to look ahead and purposely carry out actions designed to sustain the group and improve its competitive position. These features may actually represent an exhaustive list of the precise attributes of a species that would maximize its likelihood of significant group selection, or evolution by differential extinction of groups. Thus group selection involves the paradox that competing populations must be sufficiently isolated to become different in ways that may lead to their differential extinction yet close enough together that they can replace one another. This condition is obviously fulfilled with sympatric competing species, which are intrinsically isolated. So, to some extent, are hostile neighboring populations of humans. It is an important result of the above considerations that in seeking to define the adaptiveness of culture, to analyze directions of cultural change, and to identify sources of cultural rules, we cannot ignore or downplay effects significant at the group level. (Alexander 1974: 376–7)

Alexander here could be talking about genes or culture responding to group selection. In his *Darwinism and Human Affairs* (1979), he devotes a chapter to culture. Like Wilson, Alexander is reluctant to give any fundamental evolutionary role to culture and cultural evolution:

> Cultural novelties do not replicate or spread themselves, even indirectly. They are replicated as a consequence of the behavior of the vehicles of gene replication. Only if decisions or tendencies of such vehicles of gene replication (individuals) to use or not use a cultural novelty are independent of the interests of the genetic replicators can it be said that cultural change is independent of the differential reproduction of genes. (Alexander 1979: 80)

The writing of the early cultural evolutionists may not have been entirely clear at this early date; however, our thinking was not that culture was independent of genes, but that genes and culture were locked together in a coevolutionary dance with strong reciprocal interactions—the Marmaduke hypothesis.

Much of Alexander (1979) applies evolutionary theories of behaviour to anthropological issues such as kinship. For example, in matrilineal societies men are expected to care for their sister's offspring rather than their own. If certainty about paternity is sufficiently low, men will tend to be more closely related to their sister's offspring than to their own (putative) offspring. This direct address of anthropological problems by Alexander

(1971) was an important inspiration for the first generation of human behavioural ecologists (Chagnon and Irons 1979). In spite of his general explanatory strategy being similar to Wilson's, Alexander is not a fan of the human nature concept. In a short epilogue entitled 'On the Limits of Human Nature', he writes that the three reviewers of his manuscript 'were disappointed that I had not more explicitly attacked the problem of human nature, identifying its limits and explaining the consequences'. He explains why:

> As it concerns human social behavior, human nature would seem to be represented by our learning capabilities and tendencies in different situations. The limits of human nature, then, could be identified by discovering those things that we cannot learn. But there is a paradox in this, for to understand human nature would then be to know how to change it—how to create situations that would enable or cause learning that could not previously occur. To whatever extent this is so, the limits of human nature become will-o'-the-wisps that inevitably retreat ahead of our discoveries about them. (Alexander 1979: 279)

As we have seen in the case of Wilson, and perhaps also in the case of Alexander's reviewers, a purist account of human evolution based on the Modern Synthesis would lead to an account that could not do justice to human culture. It would also lead to an account that would offer no room for improving the human condition. Any concession at all to culture or reason while remaining fully committed to the Modern Synthesis leads to a paradox, as Alexander notes.

Wilson and Alexander were the first evolutionary biologists to offer sophisticated expositions on human nature. The next question is how the usages of 'human nature', and reactions to it, have evolved in the last four decades.

## 8.6 John Tooby and Leda Cosmides

Tooby and Cosmides (1990), a paper heavy in evolutionary theory, includes a rather thorough rehearsal of Modern Synthesis evolutionary theory as it applies to human evolution. They stress two principles that are foundational for their later thinking. First, they emphasize what they call 'universal functional design'. By this principle, most adaptations are tightly integrated function complexes that do not vary within a species. Functional complexity is simply intolerant of variation. Most variation that does exist is in the form of trivial, neutral differences. If this principle is true, evolution is a slow process, and adaptation to the Pleistocene would have equipped humans with a complex of adaptations for the regularities of that environment. If human cognitive adaptations are adaptations to Pleistocene life, then their nature can be deduced by what would have been adaptive in that environment. Their subsequent quarter-century-long research programme rests heavily on this principle. It comes from R. A. Fisher's (1958) picture of adaptation. Interestingly, evolutionary biology in the ensuing years has found reasons to be quite sceptical of Fisher's ideas in this regard (Thompson 2013). Selection often seems to favour adaptations to features of the environment that are quite local in time and space, as in the famous case of the evolution of the beaks of Galapagos finches in response to a succession of drought and rain (Grant and Grant 2002). It is also interesting that 'the' Pleistocene turns out to have been characterized by an enormous amount of quite noisy variation on the millennial and sub-millennial time scales. This variation seems to have increased progressively over the last few hundred thousand years. Cultural evolutionary theory suggests that this is just the kind of variation that could favour the evolution of a costly system of culture (Richerson and Boyd 2013). Perhaps culture was originally an adaptation to the unpredictable irregularities of the Pleistocene.

Tooby and Cosmides' second foundational principle is the idea of 'condition-responsive adaptive strategies'. Much variation arises because of environmentally sensitive, contingent developmental programmes or alternative genetically programmed morphs, as in the case of sexual dimorphism. Modern developmental biology has amply confirmed the existence of condition-responsive regulatory circuits active during development and even in response to environmental contingencies in adulthood (Gilbert and Eppel 2008). Ironically, some of these developmental processes include forms of gene-based Lamarckian inheritance, as well as more conventional alternative developmental pathways (Jablonka 2012). Tooby and Cosmides (1992) used this principle for a full-scale assault to the social sciences' dependence on the concept of culture, to the point of dismissing learning and culture as ideas that must die (Tooby 2015).

Tooby and Cosmides were thus the most uncompromising human nature theorists, offering no comforting notions of how we might deal with the failure of Pleistocene adaptations in a Holocene world by reason, acts of will, or scientific progress. But, in the end, they embraced the now familiar idea of human reason (Cosmides et al. 2001). In earlier work, they had championed the idea of the mind as massively modular, built of specialized, informationally encapsulated cognitive subsystems, evolved in Pleistocene environments and of limited functionality in the Holocene (never mind that we have been much more successful in the Holocene than in the Pleistocene); only systems such as these could possibly evolve. However, in their (2001), they took the contrary tack, arguing that human minds also have a powerful 'improvisational intelligence' that can account for how humans have done so well in both the Pleistocene and Holocene (see also Pinker 2010). In other words, after some lapse, they made one of the standard moves the Modern Synthesis approach to human nature must make to rescue it from a patent inability to explain the facts of the human condition or to offer a way out of the moral and political problems it poses. The cultural evolutionists make their now familiar objections: developments in technology and social organization in both the Pleistocene and the Holocene have been faster than is easy to explain by genetic evolution, but far slower than can be accounted for by powerful individual improvisational intelligence alone (Boyd et al. 2011). Cultures also show the distinctive pattern of descent with modification that are the hallmarks of Darwinian evolution (Richerson et al. 2016). Phylogenetic patterns of cultural ancestry can be traced for millennia (e.g. Currie et al. 2010).

## 8.7 Steven Pinker

Pinker's *The Blank Slate: The Modern Denial of Human Nature* (2002) is a long book that attributes to the pioneers of the social sciences a taste for biology-free, blank-slate explanations of human behaviour on the one hand, while also exhibiting a tendency towards a noble savage view of human nature on the other. This seems confused from the outset. The noble savage idea—namely, that virtue is built into human nature—seems to be at odds with a blank slate. On the Modern Synthesis view, human prosociality is hard to explain. On the noble savage view, human vice is hard to explain. Both are strong positions on human nature, with obvious problems. For example, experimental games reveal variation in prosocial behaviour at the individual, intracultural community, and intercultural levels (Henrich et al. 2004). A few humans are noble and a few are psychopaths; most of us are morally muddled to various degrees.

Pinker thinks that biology needs to be brought into the science of human behaviour via evolutionary cognitive psychology, an argument very similar to that of Tooby and Cosmides. He offers Chomsky's ideas of the innate cognitive foundations of language as a successful model (2002: 37–8).[3] Like Wilson, he discusses culture at some length, arguing, for example, that its foundation must be the cognitive architecture that humans evolved during the Pleistocene. He recognizes that life in complex societies is built upon social realities. Of the human nature theorists I discuss in this chapter, it is Pinker who makes the most use of cultural factors and processes in his explanations of human behaviour. However, he ignores the claims of cultural evolutionists that cultural evolution might have played some fundamental role in the evolution of genes, and bypasses any attempt to solve the paradox that has vexed most human nature theorists.

Pinker (2002) really only defends a very weak version of human nature, in the same vein as Wilson (1978). He rehearses the human universals argument of Brown (1991). One can make a very long list of human traits that are universal or near-universal in human societies. Some of these plausibly have biological components, such as childhood fear of strangers. But such emotions are not restricted to humans (Panksepp and Biven 2012). Other 'universals', like marriage and kinship, are culturally quite variable. The proper universal with regard to marriage is physiological in the first instance, not cognitive: the very large brain, helplessness, and slow growth of human infants means that, unlike other apes, mothers cannot raise them unaided (Burkart et al. 2009; Hrdy 2009). Humans use a great variety of marriage and kinship systems to organize the assistance that mothers need to raise their children. As we have seen, arguments of this type just point to the fact that biological factors and biological evolution are important in humans no matter how important culture and cultural evolution might also be. It is hard to believe that any educated person would believe differently. Blank-slaters are not an important part of ongoing debates. Gene–culture coevolutionists do

---

[3] Evolutionary linguists mostly now consider such highly innatist approaches to syntax dubious (Hurford 2011; Newmeyer 2004).

argue that the evolution of our big brain was driven by the advantages of having the complex culture that large brains make possible.

## 8.8 The Gene–Culture Coevolution Alternative to Human Nature

As some evolutionary biologists and evolutionary social scientists were beginning to develop ideas about human nature, other social scientists and evolutionary biologists began again to work on the problem of cultural evolution in much the same vein as Darwin. These efforts would eventually lead to a theory, much of it built on formal models akin to those of population genetics, that outlined in some detail the way genes and culture might interact in a unified evolutionary system. Currently, a fair number of researchers work on various empirical projects, some inspired by modelling work and some independent of it (Richerson and Christiansen 2013). The view of these scholars came to be that the theory of human nature developed by Wilson and his Modern Synthesis-inspired successors misrepresents the relationship between genetic and cultural evolution. Theory and some evidence suggests that for at least the last few tens of thousands of years, if not for far longer, human organic evolution was substantially driven not by natural selection acting on genes, but by culturally mandated social selection acting directly on human genes and indirectly by natural selection in human-constructed environments.

This work began early, but was a long time in reaching maturity. Gerard et al. (1956) wrote a prescient programmatic essay outlining how cultural evolution could be studied using concepts and methods similar to those used by evolutionary biologists. Donald Campbell (1965) wrote a similar essay that was more widely read. Cavalli-Sforza and Feldman (1973a, 1973b) showed how population genetic models could be modified to make models of cultural transmission and evolution. A brief mention in Richard Dawkins's *The Selfish Gene* (1976) invited readers to consider that cultural elements (memes) could evolve much like genes. Richerson and Boyd (1978) modelled the idea that even under the assumption that the human culture capacity is at a genetic fitness optimum, individual cultural traits could deviate from their genetic fitness-optimizing value if genetic mechanisms could only optimize the genetic fitness of the culture capacity as a whole, not cultural trait by cultural trait. In essence, humans might be the analogue of a two-species mutualism, the Marmaduke hypothesis, in which the interests of the two partners are fairly closely, but not perfectly, aligned.

In the first instance, the difference of opinion between the cultural evolutionists and the proponents of human nature turned on different intuitions derived from formal models. The assumptions of the models constructed by Lumsden and Wilson (1981, 2006) differed in critical details from the population genetics-style models pioneered by Cavalli-Sforza and Feldman (1973a, 1973b). The former considered that culture consists of an inventory of knowledge items that everyone knows ('culturgens'). What varies is the usage of the items, and this is guided by 'epigenetic rules': either genetically coded rules that are indifferent to usage rates (*tabula rasa* genotype) or genotypes that bias usage in favour of one culturgen or another. Lumsden and Wilson allow for cultural variation by incorporating 'trend watching', a tendency to use the same culturgen as others (in the simplest case). In models in the Cavalli-Sforza and Feldman tradition, cultural variation is assumed to represent different people having different knowledge; cultural variation is analogous to different genetic alleles. These structural differences are less important than the simplifying assumptions used in their analysis. In key parts of their analysis, Lumsden and Wilson simplify away innovation and make trend watching extremely weak. Thus, the only things that can evolve in their analysis are the gene-based epigenetic rules. Culture, in the sense represented in Cavalli-Sforza and Feldman's models, disappears due to a simplification Lumsden and Wilson apparently assume is innocent. This leads to what they call the 'thousand-year rule'. A more-fit epigenetic rule will undergo a selective sweep from low to high frequency in roughly forty human generations. As humans adapt to environments that vary in time and space, including cultural components, epigenetic rules will evolve to adapt cultural traits to them.

In these population genetics-style models of cultural evolution, innovation, migration, and errors of learning will create patterns of culturally heritable variation. If individual decision-making is a very strong effect, the heritability effect will be slight and models will collapse back to something like Lumsden and Wilson's analysis, or to the picture given by Tooby and Cosmides (Boyd and Richerson 1985). However, many cultural traditions are complex and difficult for individuals to evaluate. People will mostly have to acquire such cultural variants by learning from others. At the same time, we have general-purpose learning biases that can exert at least weak selective effects on cultural variants (Henrich 2016). Foods that are high in energy and protein are more reinforcing than ones low in the nutrients humans need. Such biases have shaped a huge variety of cuisines around the world, usually starting with ingredients that are inedible without sophisticated extraction and processing (Hill et al. 2009). To the extent that culture acts as an inheritance system, natural selection can also act directly on cultural variation. More nutritious diets will increase fitness and poor ones will decrease it, all else being equal. Furthermore, culture creates environments that can select genetic variants, including social systems that exert social selection by rewarding those who conform to cultural rules or punishing those who don't. By the early 1980s, the gene–cultural coevolution position was well articulated (Boyd and Richerson 1983; Campbell 1975; Pulliam and Dunford 1980; Richerson and Boyd 1978), and Wilson (1978: 235–6, n.) associated the term 'human nature' with his position in the debate.

In the final analysis, the issue between these two approaches is whether the evolution of genes—as extended to behaviour by Hamilton, Trivers, and Maynard Smith—provides a sufficient theoretical toolkit to explain human behaviour, or instead does the evolution of culture make a fundamental difference to the process of evolution in our species, requiring tools not part of the Modern Synthesis? My argument in this chapter is that 'human nature' became an umbrella term for the position that the Modern Synthesis provides all the evolutionary theory needed to explain human behaviour, and that culture plays no ultimate explanatory role. As with Lewens's (2012) critique of neo-Aristotelian ethics, I also argue that the version of the human nature account reviewed here can't do the work it is intended to do. The human nature theorists end up giving culture and reason strong roles to play, but without conceding a theoretically important role for culture or reason in human evolution. They variously allow that recent, contemporary, or future phases of our evolution may be strongly affected by cultural and reason; yet it is easy to imagine that deep into the Pleistocene, processes like culturally driven social selection shaped fundamental aspects of our genetically transmitted social psychology.

If stone tools and larger brains are a valid index, the species at the inception of our genus, some 2.6 million years ago, show more signs of culture than do other ape species. Brain size and stone tool complexity grow more or less in tandem throughout the Pleistocene (Richerson and Boyd 2013), as if brain enlargement was for the purpose of acquiring and managing a growing cultural repertoire. Some paleoanthropologists suggest that many of the elements of modern human behaviour appear at a regular pace over the last 250,000 years (Marean 2010; McBrearty and Brooks 2000), although equally competent authorities see a more punctuated pattern associated with the Upper Paleolithic Transition in Europe 50,000–40,000 years ago (Klein and Edgar 2002). Even this foreshortened timescale leaves a period four to five times longer than Wilson (1978) allowed for cultural evolutionary processes to affect genetic evolution. Good evidence suggests that the best-documented case of culture-driven gene–culture coevolution, the evolution of adult lactase persistence following the evolution of dairying, resulted from culture-driven selection acting on the timescale of a millennium (Itan et al. 2009)—recalling Lumsden and Wilson's (1981) thousand-year rule, but in reverse. Many other genetic changes in human populations appear to have been driven by cultural changes (Laland et al. 2010; Richerson et al. 2010). Cultural evolutionists thus argue that the basic Pleistocene hunting-and-gathering adaptation evolved under the influence of cultural evolution, including wilful processes like selective imitation of more successful groups by less successful ones, and selective migration from poorly performing to better-performing groups (Richerson et al. 2016). While we know very little about the exact trajectory of gene–culture coevolution, we know that stone tools were present at the beginning of the genus *Homo*, and that both living and Late/Upper Paleolithic hunter-gatherers are modern as regards genetic factors (Hill et al. 2009).

## 8.9 Conclusion

Lewens's (2012) analysis of the 'human nature' concept and the uses to which people have tried to put it receives support from this review of its use by sophisticated evolutionary biologists and psychologists. A weak version of human nature that points out that genes and their evolution ought to have an important part in the explanation of human behaviour should be uncontroversial. Some social scientists and humanists might have resisted, and may still resist, even this version of human nature, but on what persuasive grounds it is hard to imagine.

But the several theorists using the 'human nature' term reviewed here at least seem to want a stronger version of human nature that does real work. The strong claim is that natural selection acting on genes is the master force in human evolution, and that culture is a proximate system that has only an indirect role in evolutionary analysis (Dickins and Rahman 2012). Wilson's (1978) original formulation, echoed in several of his subsequent works, is that selection on genes is foundational in evolution, and that genes are foundational in development. Thus culture and learning, coming late in evolutionary and developmental time, play only a proximal role in explaining human behaviour. The cultural evolutionists argue that culture and intelligence were probably already present in simple forms in our ape ancestors (de Waal 2000; Whiten et al. 1999). For more than two million years, human brains and human culture evolved together (Klein 2009). Similarly, one of the earliest skills to mature in infants is the capacity for social learning (Carey 2009). Thus the strong human nature claim is undermined by the fact that genes and culture coevolve and codevelop in our species; cultural evolution is part of the ultimate explanation for human behaviour, and perhaps proximate factors have a role in evolution, *contra* Mayr (1961). The claim is further undermined by human nature theorists' need to appeal to culture and reason in various ways to repair or supplement their accounts.

How do we account for the resistance of the authors reviewed here to providing theoretical room for cultural evolution and other processes in their accounts? In the early days, I regarded cultural evolutionary theory to be a minor, friendly amendment to the orthodox Modern Synthesis in the case of one species. Some evolutionary biologists have been quite passionate in their objections to cultural evolution theory, even while showing little sign of actually understanding the cultural evolutionists' arguments (West et al. 2011). One strong thread that ties Dobzhansky to all the subsequent sophisticated biologists defending the human nature concept is a strong commitment to the Modern Synthesis, with its rejection of a fundamental role for cultural evolutionary processes in human evolution. This is often coupled with a commitment to a strong version of Ernst Mayr's (1961) proximate/ultimate distinction: that culture is a proximate mechanism and that evolutionary explanations rooted in genes furnish the only legitimate ultimate explanation. Processes like gene–culture coevolution, transgenerational epigenetic inheritance, and niche construction suggest that the Modern Synthesis needs to be replaced by an extended evolutionary synthesis (Laland et al. 2011). The espousal of a strong concept of human nature, or the denial that such is possible, is part of a deep debate in modern evolutionary biology. A whiff of grapeshot and a crackle of thin ice, indeed!

## References

Alexander, R. D. (1971). 'The Search for an Evolutionary Philosophy of Man.' *Proceedings of the Royal Society of Victoria* 84: 99–120.

Alexander, R. D. (1974). 'The Evolution of Social Behavior.' *Annual Review of Ecology and Systematics* 5: 325–83.

Alexander, R. D. (1979). *Darwinism and Human Affairs*. Seattle: University of Washington Press.
Alexander, R. D. (1987). *The Biology of Moral Systems*. Hawthorne, NY: Aldine de Gruyter.
Baum, W. M. (2005). *Understanding Behaviorism: Behavior, Culture, Evolution*. Malden, Mass.: Blackwell.
Boehm, C. (2008). 'Purposive Social Selection and the Evolution of Human Altruism.' *Cross-Cultural Research* 42: 319–52.
Boyd, R., and Richerson, J. (1983). 'Why Is Culture Adaptive?' *Quarterly Review of Biology* 58: 209–14.
Boyd, R., and Richerson, J. (1985). *Culture and the Evolutionary Process*. Chicago: University of Chicago Press.
Boyd, R., Richerson, J., and Henrich, J. (2011). 'The Cultural Niche: Why Social Learning Is Essential for Human Adaptation.' *Proceedings of the National Academy of Sciences* 108: 10918–25.
Brattain, M. (2007). 'Race, Racism, and Antiracism: UNESCO and the Politics of Presenting Science to the Postwar Public.' *American Historical Review* 112: 1386–1413.
Brown, D. E. (1991). *Human Universals*. Philadelphia: Temple University Press.
Burian, R. M. (1988). 'Challenges to the Evolutionary Synthesis.' *Evolutionary Biology* 23: 247–69.
Burkart, J. M., Hrdy, S. B., and Van Schaik, C. P. (2009). 'Cooperative Breeding and Human Cognitive Evolution.' *Evolutionary Anthropology* 18: 175–86.
Campbell, D. T. (1965). 'Variation and Selective Retention in Socio-Cultural Evolution.' In H. R. Barringer, G. I. Blanksten, and R. W. Mack (eds), *Social Change in Developing Areas: A Reinterpretation of Evolutionary Theory*, 19–49. Cambridge, Mass.: Schenkman.
Campbell, D. T. (1975). 'On the Conflicts between Biological and Social Evolution and between Psychology and Moral Tradition.' *American Psychologist* 30: 1103–26.
Carey, S. (2009). *The Origin of Concepts*. New York: Oxford University Press.
Cavalli-Sforza, L. L., and Feldman, M. W. (1973a). 'Cultural versus Biological Inheritance: Phenotypic Transmission from Parents to Children (A Theory of the Effect of Parental Phenotypes on Children's Phenotypes).' *American Journal of Human Genetics* 25: 618–37.
Cavalli-Sforza, L. L., and Feldman, M. W. (1973b). 'Models for Cultural Inheritance, I: Group Mean and Within-Group Variation.' *Theoretical Population Biology* 4: 42–55.
Chagnon, N. A., and Irons, W. (1979). *Evolutionary Biology and Human Social Behavior: An Anthropological Perspective*. North Scituate, Mass.: Duxbury Press.
Chudek, M., and Henrich, J. (2011). 'Culture–Gene Coevolution, Norm-Psychology and the Emergence of Human Prosociality.' *Trends in Cognitive Sciences* 15: 218–26.
Cosmides, L., Tooby, J., and Barkow, J. H. (2001). 'Unravelling the Enigma of Human Intelligence: Evolutionary Psychology and the Multimodular Mind.' In R. J. Sternberg and J. C. Kaufman (eds), *The Evolution of Intelligence*, 145–99. Hillsdale, NJ: Erlbaum.
Currie, T. E., Greenhill, S. J., Gray, R. D., Hasegawa, T., and Mace, R. (2010). 'Rise and Fall of Political Complexity in Island South-East Asia and the Pacific.' *Nature* 467: 801–4.
Darwin, C. (1874). *The Descent of Man, and Selection in Relation to Sex*. New York: American Home Library.
Dawkins, R. (1976). *The Selfish Gene*. Oxford: Oxford University Press.
de Waal, F. (2000). *Chimpanzee Politics: Power and Sex among Apes*. Baltimore: Johns Hopkins University Press.
Dickins, T. E., and Rahman, Q. (2012). 'The Extended Evolutionary Synthesis and the Role of Soft Inheritance in Evolution.' *Proceedings of the Royal Society B* 279: 2913–21.

Dobzhansky, T. (1962). *Mankind Evolving: The Evolution of the Human Species.* New Haven, Conn.: Yale University Press.

Dobzhansky, T., and Montagu, M. F. A. (1947). 'Natural Selection and the Mental Capacities of Mankind.' *Science* 105: 587–90.

Fisher, R. A. (1958). *The Genetical Theory of Natural Selection.* New York: Dover.

Gerard, R. W., Rapoport, A., and Kluckhohn, C. (1956). 'Biological and Cultural Evolution: Some Analogies and Explorations.' *Behavioral Science* 1: 6–34.

Ghiselin, M. T. (1997). *Metaphysics and the Origin of Species.* New York: SUNY Press.

Gilbert, S. F., and Eppel, D. (2008). *Ecological Developmental Biology: Integrating Epigenetics, Medicine, and Evolution.* Sunderland, Mass.: Sinauer.

Gould, S. J. (1982). 'Darwinism and the Expansion of Evolutionary Theory.' *Science* 216: 380–7.

Grant, R., and Grant, B. R. (2002). 'Unpredictable Evolution in a 30-Year Study of Darwin's Finches.' *Science* 296: 707–11.

Gruber, H. E., and Barrett, P. H., (1974). *Darwin on Man: A Psychological Study of Scientific Creativity Together with Darwin's Early and Unpublished Notebooks.* New York: Dutton.

Hämäläinen, P. (2008). *The Comanche Empire.* New Haven, Conn.: Yale University Press.

Hamilton, W. D. (1964). 'Genetic Evolution of Social Behavior I, II.' *Journal of Theoretical Biology* 7: 1–52.

Henrich, J. (2016). *The Secret of Our Success: How Culture Is Driving Human Evolution, Domesticating Our Species, and Making Us Smarter.* Princeton, NJ: Princeton University Press.

Henrich, J., Boyd, R., Bowles, S., Camerer, C., Fehr, E., and Gintis, H. (2004). *Foundations of Human Sociality: Economic Experiments and Ethnographic Evidence from Fifteen Small-Scale Societies.* Oxford: Oxford University Press.

Hill, K., Barton, M., and Hurtado, A. M. (2009). 'The Emergence of Human Uniqueness: Characters Underlying Behavioral Modernity.' *Evolutionary Anthropology* 18: 174–87.

Hodgson, G. M. (2004). *The Evolution of Institutional Economics: Agency, Structure and Darwinism in American Institutionalism.* London: Routledge.

Hrdy, S. B. (2009). *Mothers and Others: The Evolutionary Origins of Mutual Understanding.* Cambridge, Mass.: Harvard University Press.

Hull, D. L. (1986). 'On Human Nature.' *Proceedings of the Biennial Meeting of the Philosophy of Science Association*: 3–13.

Hurford, J. R. (2011). *The Origins of Grammar: Language in the Light of Evolution II.* Oxford: Oxford University Press.

Huxley, J. (1942). *Evolution: The Modern Synthesis.* New York: Harper.

Itan, Y., Powell, A., Beaumont, M. A., Burger, J., and Thomas, M. G. (2009). 'The Origins of Lactase Persistence in Europe.' *PLoS Computational Biology* 5: e1000491.

Jablonka, E. (2012). 'Epigenetic Inheritance and Plasticity: The Responsive Germline.' *Progress in Biophysics and Molecular Biology* 111: 99–107.

Jensen, A. S. (1969). 'How Much Can We Boost IQ and Scholastic Achievement?' *Harvard Educational Review* 39: 1–123.

Jorgensen, J. G. (1980). *Western Indians: Comparative Environments, Languages, and Cultures of 172 Western American Indian Tribes.* San Francisco, Calif.: W. H. Freeman.

Klein, R. G. (2009). *The Human Career: Human Biological and Cultural Origins.* Chicago: University of Chicago Press.

Klein, R. G., and Edgar, B. (2002). *The Dawn of Human Culture: A Bold New Theory on What Sparked the 'Big Bang' of Human Consciousness.* New York: John Wiley.

Kroeber, A. L. (1948). *Anthropology: Culture Patterns and Processes*. New York: Harcourt, Brace & World.

Krubitzer, L. A., and Seelke, A. M. H. (2012). 'Cortical Evolution in Mammals: The Bane and Beauty of Phenotypic Variability.' *Proceedings of the National Academy of Sciences* 109: 10647–54.

Laland, K. N., Odling-Smee, J., and Myles, S. (2010). 'How Culture Shaped the Human Genome: Bringing Genetics and the Human Sciences Together.' *Nature Reviews Genetics* 11: 137–48.

Laland, K. N., Sterelny, K., Odling-Smee, J., Hoppitt, W., and Uller, T. (2011). 'Cause and Effect in Biology Revisited: Is Mayr's Proximate–Ultimate Dichotomy Still Useful?' *Science* 334: 1512–16.

Lewens, T. (2012). 'Human Nature: The Very Idea.' *Philosophy and Technology* 25: 459–74.

Lumsden, C. J., and Wilson, E. O. (1981). *Genes, Mind, and Culture: The Coevolutionary Process*. Cambridge, Mass.: Harvard University Press.

Lumsden, C. J., and Wilson, E. O. (2006). *Genes, Mind, and Culture: The Coevolutionary Process*. Hackensack, NJ: World Scientific.

Marean, C. W. (2010). 'Pinnacle Point Cave 13B (Western Cape Province, South Africa) in Context: The Cape Floral Kingdom, Shellfish, and Modern Human Origins.' *Journal of Human Evolution* 59: 425–43.

Mayr, E. (1961). 'Cause and Effect in Biology.' *Science* 134: 1501–6.

McBrearty, S., and Brooks, A. S. (2000). 'The Revolution That Wasn't: A New Interpretation of the Origin of Modern Human Behavior.' *Journal of Human Evolution* 39: 453–563.

Mesoudi, A., Blanchet, S., Charmantier, A., et al. (2013). 'Is Non-Genetic Inheritance Just a Proximate Mechanism? A Corroboration of the Extended Evolutionary Synthesis.' *Biological Theory* 7: 189–95.

Morgan, T. J. H. (2016). 'Testing the Cognitive and Cultural Niche Theories of Human Evolution.' *Current Anthropology* 57: 370–77.

Newmeyer, F. J. (2004). 'Against a Parameter-Setting Approach to Typological Variation.' *Linguistic Variation Yearbook* 4: 181–234.

Panksepp, J., and Biven, L. (2012). *The Archaeology of Mind: Neuroevolutionary Origins of Human Emotions*. New York: W. W. Norton.

Pinker, S. (2002). *The Blank Slate: The Modern Denial of Human Nature*. New York: Penguin.

Pinker, S. (2010). 'The Cognitive Niche: Coevolution of Intelligence, Sociality, and Language.' *Proceedings of the National Academy of Sciences* 107: 8993–9.

Pulliam, H. R., and Dunford, C. (1980). *Programmed to Learn: An Essay on the Evolution of Culture*. New York: Columbia University Press.

Richards, R. J. (1987). *Darwin and the Emergence of Evolutionary Theories of Mind and Behavior*. Chicago: University of Chicago Press.

Richerson, P., Baldini, R., Bell, A., et al. (2016). 'Cultural Group Selection Plays an Essential Role in Explaining Human Cooperation: A Sketch of the Evidence, together with Commentaries and Authors' Response.' *Behavioral and Brain Sciences* 39: 1–68.

Richerson, J., and Boyd, R. (1978). 'A Dual Inheritance Model of the Human Evolutionary Process, I: Basic Postulates and a Simple Model.' *Journal of Social Biological Structures* 1: 127–54.

Richerson, J., and Boyd, R. (2005). *Not by Genes Alone: How Culture Transformed Human Evolution*. Chicago: University of Chicago Press.

Richerson, J., and Boyd, R. (2013). 'Rethinking Paleoanthropology: A World Queerer Than We Supposed.' In G. Hatfield and H. Pittman (eds), *Evolution of Mind, Brain, and Culture*, 263–302. Philadelphia: University of Pennsylvania Museum of Archaeology and Anthropology.

Richerson, J., Boyd, R., and Henrich, J. (2010). 'Gene–Culture Coevolution in the Age of Genomics.' *Proceedings of the National Academy of Science of the USA* 107: 8985–92.

Richerson, J., and Christiansen, M. H. (2013). *Cultural Evolution: Society, Technology, Language, and Religion*. Cambridge, Mass.: MIT Press.

Ross, C. T., and Richerson, J. (2014). 'New Frontiers in the Study of Cultural and Genetic Evolution.' *Current Opinion in Genetics and Development* 29: 103–9.

Schwitalla, A. W., Jones, T. L., Pilloud, M. A., Codding, B. F., and Wiberg, R. S. (2014). 'Violence among Foragers: The Bioarchaeological Record from Central California.' *Journal of Anthropological Archaeology* 33: 66–83.

Segerstrale, U. (2000). *Defenders of the Truth: The Battle for Science in the Sociobiology Debate and Beyond*. New York: Oxford University Press.

Smith, J. M. (1976). 'Evolution and the Theory of Games: In Situations Characterized by Conflict of Interest, the Best Strategy to Adopt Depends on What Others Are Doing.' *American Scientist* 64: 41–5.

Thompson, J. N. (2013). *Relentless Evolution*. Chicago: University of Chicago Press.

Tooby, J. (2015). 'Learning and Culture.' In J. Brockman (ed.), *This Idea Must Die: Scientific Theories That Are Blocking Progress*, 432–6. New York: Harper.

Tooby, J., and Cosmides, L. (1990). 'On the Universality of Human Nature and the Uniqueness of the Individual: The Role of Genetics and Adaptation.' *Journal of Personality* 58: 17–67.

Tooby, J., and Cosmides, L. (1992). 'The Psychological Foundations of Culture.' In J. Barkow, L. Cosmides, and J. Tooby (eds), *The Adapted Mind: Evolutionary Psychology and the Generation of Culture*, 19–136. New York: Oxford University Press.

Tostevin, G. B. (2013). *Seeing Lithics: A Middle-Range Theory for Testing Cultural Transmission in the Pleistocene*. Oxford: Harvard University American School of Prehistoric Research and Oxbow Books.

Trivers, R. L. (1971). 'The Evolution of Reciprocal Altruism.' *Quarterly Review of Biology* 46: 35–57.

West, S. A., El Mouden, C., and Gardner, A. (2011). 'Sixteen Common Misconceptions about the Evolution of Cooperation in Humans.' *Evolution and Human Behavior* 32: 231–62.

Whiten, A., Goodall, J., McGrew, W. C., et al. (1999). 'Cultures in Chimpanzees.' *Nature* 399: 682–5.

Williams, G. C. (1966). *Adaptation and Natural Selection: A Critique of Some Current Evolutionary Thought*. Princeton, NJ: Princeton University Press.

Wilson, E. O. (1975). *Sociobiology: The New Synthesis*. Cambridge, Mass.: Harvard University Press.

Wilson, E. O. (1978). *On Human Nature*. Cambridge, Mass.: Harvard University Press.

# 9

# Human Ontogenies as Historical Processes
An Anthropological Perspective

*Christina Toren*

## 9.1 Introduction

Across the human sciences one finds theoretical perspectives that recognize the nature/culture distinction as untenable. At the same time, the gap between demonstrating its inadequacy and developing a viable alternative approach is wide indeed. I have long argued that the study of human ontogeny as a microhistorical process frees the analyst from the conceptual impasse produced by the analytical distinction between biology and culture (ditto for society/individual, *langue/parole*, structure/process, and so on). My various studies of the ideas and practices of villagers in rural Fiji have shown that whatever the focus of investigation, children of different ages and adults have ideas about the world and themselves that are at once unique to the particular person and vary systematically as a function of age and of their lived history as Fijians. So, for example, their descriptions of the same ritual gathering, while manifestly versions of one another, are at the same time systematically at odds with one another, such that it becomes evident that adult ideas of meaningful practice stand as ground to each child's own understanding considered as figure (see e.g. Toren 1990, 1993). Such variation suggests the endogenous transformation of meaning over time—which, after all, any contemporary theory of learning from Piaget onwards (however constrained, or not, by notions of certain forms of conceptual apparatus as 'innate' or 'hard-wired') would lead us to expect. The present chapter provides a way of thinking about historical processes that allows us to address what people say and do and think in their own terms—that is to say, in terms of how they actually make sense of the world and themselves.

## 9.2 Historical Processes

I begin with the issue of how we conceive of historical processes. The idea of history as lived is derived from Merleau-Ponty, who writes that he had it from Proust. My usage is not, therefore, quite idiosyncratic. I take it to be axiomatic that each one of us continues through time, from conception to death, as the transforming autopoietic—that is, self-creating, self-producing—product of an always unique history of social relations in and through which we think and speak and act; human autopoiesis is through and through a social process such that we become who we are as a function of our relations with others. This history that is uniquely our own encompasses physiological and molecular processes, which are themselves amenable to analysis as self-producing systems whose continuity through time is a function of transformation. I emphasize, too, that our lived histories take in those of previous generations insofar as their histories became ours as a function at once of the specific molecular constitution of the initial zygote (including the DNA and everything else in it), with its intrinsic self-organizing tendencies; of how we were parented as a function of how our parents were parented and their parents, back through many generations; and of our encountering those others whose histories thereby intersect our own—more or less significantly.

This history of social relations that we live evinces itself in all and every aspect of our environing world, and encompasses everything about us from our physiology to our most private thoughts. It is perhaps easiest to understand if we think just of those by whom we were parented (they may not have been our parents) and reflect on how they have informed who we are for good or ill, how we rejected or embraced their contribution to the making of us. It seems at long last to be accepted that, on the one hand, a lived history of rejection and neglect has effects that continue over generations and, at the other end of the spectrum, that a lived history of privilege does likewise.[1] My idea of 'lived history' is intended, however, to hold for all of us, irrespective of how very unremarkable, how 'normal' as it were, our particular personal histories may be. I argue that in literally every aspect of our being, we evince the always transforming history of relations with others that continues making us who we are. For each one of us, the lived present emerges out of a unique lived past and presses on into a unique lived future.

I am addressing this issue of history in an attempt to confront a helpful critic's questioning observations as follows: 'Why do you stress historical processes so much? Who or what are you arguing against? Surely everyone knows that humans have history. When biologists and others talk about 'cultural transmission' it is precisely *human history* that they are talking about—that is, the way ideas and artefacts change over time'.[2]

My answer is, first, that these observations conflate the idea of 'culture' with that of 'history', which is not my intention. Rather, I take the radical view that one can conceive of all aspects of the world—including, crucially, all dimensions of human being, indeed of all living things, as historically constituted. This perspective does away with ideas of 'human nature' and 'culture' as analytical categories, but it does not entail any denial of the science of biology and its ever more remarkable technological advances;

---

[1] See e.g. Blanden et al. (2005), Blanden and Gibbons (2006), Corak (2013), Devin (2009), and Stewart (2016).

[2] I am indebted to here to Murillo Pagnotta—friend, colleague, and interlocutor.

rather, it suggests the necessity for recognizing how important it is to understand how the transforming science of biology contributes to structuring our peopled world. My critic observed further that some, if not all, of his human sciences colleagues 'cannot help but make "historically constituted" identical with "culturally constituted" and thus maintain the nature/nurture distinction'—a matter I return to below. And yes, it is the case that for all I would not dream of denying evolution, I am sceptical as to the validity of the dominant gene-centred, adaptationist perspective, and I do indeed take history to be continuous with evolution: they refer to the same process understood in terms of different timescales (cf. Robertson 1996). We are nowadays in a position to investigate the origin and development (that is to say, the ontogeny) of living things as part of complex dynamic systems, and these investigations are having a significant impact on contemporary revisions of evolutionary theory.[3]

## 9.3 Why We Need a Unified Model of Human Being

Let us at least entertain the idea that all living things evince the history of their ontogeny and that, at least in principle, it is possible for anthropologists to undertake an ethnographic analysis of this process in human beings.[4] Why do I argue that to do so we need to develop a unified model of human being? To illustrate this requires a detour to consider briefly just one aspect of Nikolas Rose's (2007) influential book *The Politics of Life Itself: Biomedicine, Power, and Subjectivity in the Twenty-First Century*. His first chapter would seem to suggest that 'biovalue'—that is, 'the value to be extracted from the vital properties of living processes'—is already so taken-for-granted an idea that the OECD report of 2004 is able, in effect, to define 'the bioeconomy' as 'bioeconomic circuits of exchange [that] have as their organizing principle the capturing of the latent value in biological processes, a value that is simultaneously that of human health and that of economic growth' (Rose 2007: 32–3). The bioeconomic circuits of exchange to which he refers are primarily those concerned with health and medical technologies, and governance of the bioeconomy is 'characterized by novel alliances between political authorities and promissory capitalism' (p. 34) that are forging 'contemporary economies of vitality' (p. 38). It follows that 'biomedical technique has extended choice to the very fabric of vital existence, we are faced with the task of deliberating about the worth of different human lives' (p. 40). Thus, the etho-politics to which Rose refers at the outset of his book—i.e. 'disputes over the value accorded to life itself' (p. 27)—were more than a decade ago already shaping a present where, as he goes on to show, the various 'forms of life' provide for the 'emergence of the living biological body as a key site for the government of individuals' (p. 255; see also Miller and Rose 2008).

---

[3] See e.g. the collection of papers edited by Gontier et al. (2006).

[4] In other words, by virtue of investigating ontogeny in specific cases as a microhistorical process, anthropologists will find themselves able to demonstrate how certain key ideas and practices are at once transformed and maintained over generations; with access to a good archive, they might even be able to use their understanding to better effect in respect of historical analysis.

Rose's argument against this perspective is certainly convincing, and I wish to take it further. From my point of view as an anthropologist who has long been concerned to achieve a new understanding of what it is to be human, the bioeconomy and the very notion of biovalue take for granted an analytical distinction between biology and culture that justifies Western scientists in general in their view of biology as a separable domain of investigation, one that gives access to knowledge of what is 'fundamental', 'natural', and 'true'.[5] International agencies such as the OECD, NGOs, and multinational corporations likewise operate in terms of this same unexamined distinction—so does the academy, by and large, and so do all its research-funding bodies.[6] Indeed, it appears that the general public (at least in the UK, North America, and other parts of the English-speaking world) take it for granted that biology's conceptual concomitant—'culture'—manifestly accounts for what is variable and, as such, prone to error; and because culture is the domain of error, it is also understood to be open to correction.[7]

Certain anthropologists have questioned how this biology/culture distinction operates. Thus Rabinow (1996) introduced the idea of 'bio-sociality' as a conceptual transformation in which the biological is transformed into the cultural by virtue of the way that the new genetics provide for what is seen as 'artificial', and Strathern (1992a, 1992b) showed how, once nature has become amenable to consumer choice, biology as the domain of explanation is thrown into question. Indeed, Sandra Bamford argued further that 'what we are confronted with [...] is a world in which biology has become unmoored' (2007: 5).[8] More recently, Descola (2013) has made an attempt to go 'beyond nature and culture'—an attempt that was bound to fail because, as I have shown elsewhere (Toren 2014), the nature/culture distinction is basic to his characterization of the schemas that structure experience. Likewise, Geoffrey Lloyd points out that both Descola and Viveiros de Castro 'challenge the nature/culture dichotomy. Yet in suggesting how to go beyond it, both do so in terms that continue to use it' (2015: 20). What seems crucial for the present chapter, however, is that the rise of biology as science and the technologies it promises 'reshape vitality from the inside: in the process

---

[5] Rose's argument builds on the work of anthropologist Sarah Franklin, who in 1995 had pointed out the close relation between the modern idea of life and the rise of the life sciences; she contributed seminal texts to the debate on the relation between 'nature' and 'culture' and contemporary bioethics (e.g. Franklin 2000, 2003).

[6] The avowed mission of the Organisation for Economic Co-operation and Development is to 'help governments tackle the economic, social and governance challenges of a globalised economy'; see www.oecd.org/.

[7] Cf. Ingold (2006: 276): 'The implied essentialisation of biology as a constant of human being, and of culture as its variable and interactive complement [...] is the single major stumbling block that [...] has prevented us from moving towards an understanding of our human selves, and of our place in the living world, that does not endlessly recycle the polarities, paradoxes and prejudices of Western thought.' One might, of course, argue that this overstates the case, that biologists are very well aware of variation and error as given in biological processes; but this awareness has not, it seems, led to any concerted attempt to get rid of the nature/culture distinction that dominates public understanding in the West.

[8] Bamford's ethnography of the Kamea of Papua New Guinea addresses 'many unexamined tenets of a biological framework' (2007: 6), including the idea that bodies exist as autonomous entities.

the human becomes, not less biological, but *all the more* biological' (Rose 2007: 19). In this process whereby biological innovations seem to promise ever more effective technologies of control (primarily in respect of medicine and gene therapies), biology as a discipline appears to become ever more explanatory, and the other human sciences, particularly the social sciences, ever less so.

To caricature the situation somewhat, but only somewhat, the mutability of the biology of living things is understood to be a function of the ever-increasing depth of our rational, scientific knowledge of biological processes. By the same token, culture comes to be seen as an ever-thinner, ever-less explanatory layer on top of the fundamentals given by biology, and 'human nature' comes to outweigh or perhaps to encompass 'culture'. Even so, the importance of the social dimension of human lives is becoming increasingly evident—for example, in respect of the treatment of medical conditions ranging from diabetes to dementia and in studies of manifold forms of learning. Theorists as various as Tomasello (2003) on language acquisition, Enfield and Levinson (2006) on sociality, Evans (2006) on class and identity, and Fogel et al. (2006) and Fogel and DeKoeyer-Laros (2007) on the development of intersubjectivity in infancy argue that learning is social through and through.[9] These apparently conflicting findings suggest that it really is time that the academy (at least) discards the distinction between biology and culture as separable domains and looks to the development of a unified model of human being. Indeed, across the human sciences, one can find theoretical perspectives that, having rendered untenable the mind/body distinction, have begun to recognize that understanding how humans make sense of the world requires a theory that is able to approach cognition as a function of the human being in relation to others in the environing world. In this view, mind cannot be a function of the brain, nor even of the embodied nervous system, and cognition cannot be a matter of bringing into play a set of representations in the brain.

## 9.4 Autopoiesis as a Historical Process

Perhaps the best-developed non-representational model of cognition is that put forward by ecological psychologist and philosopher of mind Anthony Chemero (2009), who argues for 'radical embodied cognitive science'. He has developed Gibson's theory of affordances to show not only that perception is direct—that is to say, perception does not require internal representations of the external world in the brain of the perceiving organism—but also that 'affordances [...] are relations between particular aspects of animals and particular aspects of situations' (2009: 139). Indeed, he goes further, not only taking affordances to be 'features of whole situations' but also stating, rather more specifically, that 'affordances are relations'. In this view, what we perceive

---

[9] Even those such as Meltzoff and colleagues, who are wedded to the idea of learning as a computational process, have arrived at the view that 'learning is social' (Meltzoff et al. 2009: 285). See Ingold (2007b) for a succinct account of 'the social child'.

is not in the environment alone, but is instead 'the relation between the perceiver and the environment' (p. 141). There is plenty of conceptual space here for my idea of ontogeny as a microhistorical process.

From my point of view, this theory of perception and, more broadly, of cognition is truly useful because 'the ontology of ecological psychology is not a simple form of realism. It is a form of realism about the world as it is perceived and experienced—affordances, which are inherently meaningful, are in the world, and not merely projected onto it by animals' (Chemero 2009: 151).

Compare Thompson's (2007: 36) observation that 'generative phenomenology brings to the fore the intersubjective, social, and cultural aspects of our radical embodiment [...] Individual subjectivity is [...] intersubjectivity, originally engaged with and altered by others in specific geological and cultural environments.' Leaving aside Thompson's apparent assumption that what is 'social and cultural' can be separated from what is intersubjective (to which, of course, I take exception), the key word here is 'generative'; I have been making an argument in very similar terms at least since 1999—indeed, I could say since 1993—but always in the past as something of a lone voice. Now I can take comfort in the existence of ecological psychologists such as Anthony Chemero, phenomenological biologists like Evan Thompson, and those who take the perspective of developmental systems such as Susan Oyama. Take Oyama's (2011: 85) 'developmental-systems reformulation of nature and nurture [as, respectively,] the product and process of ontogeny, [which removes] the possibility of treating [nature and nurture] as contrasting terms'. Oyama further points out that Thompson has proposed 'dismantling the opposition between self-organization and natural selection in evolution: these, too, he asserts, can be seen, not as opposing forces, but as process and product to each other.'

Further, combining Thompson's (2007) phenomenological biology—or, more exactly, the enactivist approach it entails—with Chemero's elaborated theory of affordances 'makes radical embodied cognitive science a fully dynamical science of the entire brain–body–environment system: nonrepresentational neurodynamic studies of the nervous system and sensorimotor abilities [...] match up with ecological psychological studies of affordances and sensorimotor abilities' (Chemero 2009: 154).[10] I argue that they can also match up with ethnographic studies of ontogeny as a microhistorical process and with my unified model of human being.

In achieving this objective, we can discard the biology/culture distinction and demonstrate the theoretical utility of a paradigm shift whereby the human scientist becomes capable not merely of acknowledging the lived reality of ontologies other than his or her own but of showing how other ontologies and the epistemologies that provide for them have a purchase on the transforming world that confirms their lived

---

[10] Perhaps the best-developed positions are those put forward by psychologists Oyama (2000, 2002, 2010), Lickliter (2007, 2011), Fogel et al. (2006), Fogel and and DeKoeyer-Laros (2007), and Fogel (2011), who argue for developmental systems theory.

reality (Toren 2009, 2012).[11] The fullness of the world is such that it is capable of confirming all our ideas of it, even while, often enough, we may question whether indeed we are right.[12] It follows that there is a place here for ethnographic study of the ontogenetic processes in and through which ideas of 'nature' and 'culture' come to be taken for granted as manifest in the world—so much so that we have little or no problem in foisting them on others. Here I remind the reader in passing of Strathern's 1981 paper showing that Hageners make no such distinction; nor do the rural indigenous Fijians with whom I do fieldwork. Indeed, throughout the Pacific region, the cosmologies of many (perhaps not all) indigenous peoples, including Australian aborigines, entail the idea that animals and plants and people share their substance with the land that produces them.[13]

What constitutes confirmation for any one of us is itself an artefact of our lived histories. The confirmation of a drug's efficacy that is required by a double-blind experiment differs from the confirmation that is discovered through the use of a herbal remedy where the placebo effect is not ruled out. And they both diverge markedly from the confirmation of a suspicion that ensues when new information comes to light about 'what was really going on' when a certain political deal was arrived at, or from our projections onto another person in which we find confirmation in their every action of our own ideas of their moral worth. This last point is rarely acknowledged, but is fundamental to any genuinely comparative study of human beings: in one way or another, the world by and large confirms as valid all our ideas and practices. The problem here is that in a certain view of science, the only valid explanation is one derived from hypothesis testing. This is problematic because, while experimental findings can be tremendously useful, excellent for certain kinds of investigation, other kinds of study are also useful. I argue that the scientific pursuit of valid explanations of the manifold phenomena of human lives demands that we think in terms of the unified model that I propose, precisely because it enables (indeed, requires) that we take in from the outset the dynamic complexity of human beings and their relations with one another in the world they live.

This is, however, difficult to do—so much so that for certain kinds of research it seems obvious that a factorial approach is what is wanted and most productive. So, for example, the linguistic anthropologist Eve Danziger rightly criticizes a cognitivist

---

[11] Cf. Lloyd (2015: 5): 'recognizing that reality is multidimensional allows for a plurality of accounts, each dealing with a different aspect or dimension of the subject-matter, thereby bypassing the usual dilemma that insists on a choice between "realism" and "relativism".' I argue further that because people's ideas of what constitutes the confirmation of an idea themselves vary, it is always possible to find confirmation, even in situations where those of another persuasion are just as certain of their own contrary idea.

[12] Cf. Pina Cabral (2014: 57): 'The world is one because, in personhood, alterity is anterior; as phenomenology has taught us, human experience is social before it is rational. The world, in all of its plurality, cannot escape from history; all those untold historical determinations that accumulate in the single act of any singular person.'

[13] Compare the indigenous Fijian idea that just like the earth itself and the sea, people are themselves the very substance of their places of origin, their *vanua* (land, place, country) (see Toren 1998; see also Mallett 2003, Hess 2009, Tui Atua 2009).

approach to understanding the representation of spatial orientation in language. She argues that four frames of spatial reference are to be found in use across the world—absolute, relative, intrinsic, and direct—but that not all languages make use of all of them in every context of spatial reference (Danziger 2010). The frame of spatial reference that dominates in a given language has a demonstrable effect on its speakers. To help you understand this, imagine you are looking at an array of two objects on a table—say, a kettle and a milk carton. In respect of the cardinal points of the compass, you are seated facing directly north and the milk is, from your perspective, to the right of the kettle. When asked to describe where the milk is placed in relation to the kettle, people whose language employs an absolute (but not a relative) frame of reference (FoR) reply that the milk is to the east of the kettle (e.g. speakers of Tzeltal [Mayan] or Arrernte [Pama Nyungan]); by contrast, speakers of languages that employ a relative (but not absolute) FoR reply that the milk is to the right of the kettle (e.g. speakers of Dutch or Japanese):

Speakers of [Tzeltal or Arrernte] will, if rotated 180 degrees and asked to rebuild the scene, place the milk to the east of the kettle—thus violating the original relation of the array to their own right and left sides. Speakers of languages […] in which a Relative but not an Absolute spatial frame of reference is in everyday linguistic use to describe this kind of scene, will, under the same rotation circumstances preserve the original right-left relations in their rebuilding of the scene and violate the original north-south-east-west relations in order to do so. (Danziger 2010: 174)

Danziger's work shows that pointing gestures combined with deictic demonstratives—for example, 'over there'—have to be included if we are to understand how FoR works; she further observes that 'arguments from language acquisition, cross-linguistic distribution and gesture transposition all suggest that the Direct Frame of Reference [has] a degree of psychological primacy over the other Frames' (2010: 180).

Danziger's analysis is compelling, genuinely revealing, and useful. Even so, this kind of factorial analysis, fascinating though it is, gives us only a partial grasp of how people orientate themselves in the world. It is not concerned to capture the lived understanding of space-time that evinces itself in people's relations with one another, or in how orientation skills are constituted over time as children grow up; nor does it address how the development of these same skills may be so embedded in day-to-day practices, so taken for granted, that they may never actually be discussed or even commented on. Nor is there here any consideration of the idea that how people orientate themselves and find their way around in the world may be tied in to a cosmology in which, for example, space-time orientation and way-finding validate one's very being.

In this connection, see Jennifer Biddle's paper, subtitled 'Skin as Country in the Central Desert', in which she shows how 'a certain haptic sensibility works to align certain contemporary Aboriginal bodies akin with that of the Ancestors. One does not, for instance, "tell" […] a Dreaming narrative, in Walpiri. The verb form is *yimi-purrami*, "to follow", implying that the speaker is not so much "telling" but literally "following" an already established account. *Yapa* [the Walpiri name for themselves] travel the same regions, camp in the same places, perform, where they can, the same activities as Ancestors themselves' (Biddle 2001).

Biddle's thoroughgoing analysis of Walpiri women's *yawulyu* ('ceremonial Dreaming ritual', 'ceremonial body inscriptions') shows how 'a certain embodied expression of Ancestral presence is effected: one "becomes" ancestor, "becomes" country [...] the rendering of the body commensurate with country is not a one-way process. For the aim of *yawulyu* is, after all, care of the country [...] rejuvenating, revitalizing, "feeding" certain places, species and persons' (2001: 185).

Walpiri, like Arrernte, belongs to the Pama Nyungan language family. It is instructive, therefore, to think about what kind of lived difference it might make to be orientating oneself in terms of a space-time of intersubjective relations that takes in ancestors, as well as other species and persons and the very country itself. I have some sense of this in regard to how Fijian village children arrive at an idea of themselves as future adults implicated in the historical succession of generations. The ancestors do not make any explicit appearance in the stories children wrote for me imagining their future lives, but their continuing presence is suggested in the emphasis placed by older children on their achievement of a proper status according to the land and on their continuing observation as old people of Fijian custom. The idea of the self as implicated in the historical succession of generations has rather different connotations in Fijian than it might have in English because, as I pointed out in my study:

In the Fijian language the future is not conceived of as up ahead and facing us (as it is in English usage). A literal translation of 'the past'—*na gauna eliu*—is 'the time ahead'; likewise 'the future'—*na gauna mai muri* is 'the time behind'. One's life trajectory leads one ever deeper into the past as the future unfolds behind one. This makes good sense of the succession of generations: the old move ever further forward into the past and the youngest, who are perhaps three generations behind them on this same trajectory, have at their own backs the generations to come. (Toren 2011: 27)

In examining how different understandings of intersubjective space-time figured in the ideas of 75 rural Fijian children aged between 7 and 14 years old, the paper also showed that intersubjectivity has to be regarded as historically constituted; in other words, ideas of the person and sociality inevitably structure the nature of intersubjectivity in any given case.[14] Crucial to this observation were, and are, concomitant ontogenies of ideas of the person and kinship that I have examined elsewhere.[15]

---

[14] Compare Danziger's (2013) analysis for a different perspective on Mopan Maya intersubjectivity.
[15] Lepowsky (2011) provides a fascinating account of mutually implicating ideas of self, social relations, and world in Vanatinai, PNG, in the context of Hollan and Throop's phenomenological exploration of empathy in the Pacific. The studies they bring together suggest that 'empathy emerges within an intersubjective field [...] in which self–other orientations may be differentially marked, elaborated or suppressed' (2011: 18). The collection includes an insightful chapter by Mageo in which she argues that empathy has its developmental origins in the visceral relations she characterises as 'skinship and gazeship', which lead to attachment and that 'in more socially oriented locales [such as Samoa], attachment leads to empathy as enacted', rather than to empathy as 'an inner state in which one imagines another's inner state' (2011: 76).

## 9.5 Ontogeny as a Historical Process

One may at least try to imagine how, over time, certain children are making sense, intersubjectively, of the lived space-time of places whose substance is mysteriously one with their own, where ancestral beings continue to dwell and to make themselves known, and contrast this with how other children are likewise making sense of places that (an empiricist, scientific view might suggest) can ultimately be fully known, exhausted by knowledge. The affordances the world makes available to children in such contrasting cases are historically constituted, contributing to their divergent ontogenies in respect of their constitution over time of concomitant ideas of sociality, personhood, kinship, political economy, religion, and life. Or, to put it another way, because they are aspects of the world that emerge in the relation between an organism and features of its current situation, affordances are 'out there' for us to the extent that our particular ontogenetic history has made them available.

I have argued elsewhere (Toren 2005) that ideas of what language *is good for* are bound to be intrinsic to the process of language acquisition, and thus have everything to do with how we understand, experience, and represent in speech and in writing, the environing world. It seems to me, therefore, that to grow up with an idea of language as analytical—a mode of (as it were) self-evident knowledge—is to live with reason at the forefront of existence. People like us (you, the reader, and me, the writer) live our lives in the conviction that it is possible to *know* by means of rational endeavour—even if only within the confines of our own discipline—and no doubt our children are bound to make sense of our convictions. It may be tempting to suppose that all children, everywhere, have this experience in some sense; but if so, it is bound to be exacerbated by a mode of intentionality like yours and mine, which insists on reason as given in the stuff of the world, in the sense that this stuff can be scientifically explained.

To recap: my argument is that in order to understand ontogeny, we should be rethinking our models of human being so as to have as our focus a phenomenon that is, through and through, at once historical and material, where 'mind is a function of the whole person that is constituted over time in intersubjective relations with others in the environing world' (Toren 1999b: 12).[16] I am well aware that this formulation is somewhat unwieldy, but it has seemed necessary to me to arrive at a non-representational model whose starting assumptions make the microhistorical nature of human development immediately apparent. My ethnographic studies of ontogeny have provided good evidence for my model (see e.g. Toren 1990, 1993, 1999a, 2004, 2007, 2011) and, I would argue, demonstrate its necessity. It has been pointed out to me by my helpful

---

[16] My first formulation of this model (Toren 1999b: 1–21; see also Toren 2002) was derived from a synthesis of the works of Maturana and Varela (1980, 1988) on autopoiesis; Piaget on genetic epistemology, especially his idea of the cognitive scheme as a 'self-regulating transformational system' (i.e. as an autopoietic process); Husserl's (1965, 1970) and Merleau-Ponty's (1962) phenomenology; and certain of Vygotsky's (1986) insights on language acquisition. My model has a good deal in common with Thompson (2007), and finds new support in his ground-breaking neurophenomenology; with Tomasello's (2003) work on the ontogeny of language; with Susan Oyama's (2000, 2002, 2010) work on developmental systems; and with Chemero's (2009) 'radical embodied cognitive science'.

critic that while the dominant view of the brain as an information-processing system seems to suggest that it would be easy and unproblematic, in principle, to translate psychological into neurophysiological terms, in my model it is not easy to see what role the brain is playing in the body. Point taken. Chemero's theory and the thoroughgoing experimental work that substantiates it provide what I need. To take one example, Chemero and his colleagues have been studying the dynamics of interpersonal coordination in joint action; they find that between-person structures (what they call 'interpersonal synergies') spontaneously emerge in joint action, and both constrain and enable the behaviour of each person involved. In effect, they temporarily become a coherent single unit (e.g. Walton et al. 2015). This finding is at a behavioural rather than ontogenetic timescale, but it supports my own model. Even at these short timescales, it is clear that super-personal structures partly determine the activities of the people that make up those super-personal structures. In Chemero's model and in Thompson's, the brain coordinates perception and action not because it is manipulating representations but because, as my helpful critic puts it, 'the intrinsic dynamics of the brain, itself a function of its previous developmental history, makes it particularly sensitive to certain perturbations at the sensory surfaces. These in turn lead to changes in the activity of neuromuscular structures, thus changing the spatial relations among muscles, bones, and the rest of the body attached to them.'[17] In my terms, the changing structure of the brain is a function of its embodiment in particular living persons who evince in what they do and say the history of their intersubjective relations with others in the world they live. For the sake of clarity, I reiterate here that this history of intersubjectivity is present in all our perceptions, actions, and thoughts—even the most private.

I trust it will be apparent that there is no place here for a distinction between 'biology' and 'culture'.[18] Rather, an anthropological perspective on the finding that 'affordances are relations'—in Chemero's terms, between the abilities of the organism and features of the current situation it is perceiving—cries out for ethnographic analysis of the microhistorical ontogenetic processes in and through which certain affordances are constituted.

This process through which we make sense of others, ourselves, and the environing world is intersubjectivity. The term is a shorthand for how each and every time we encounter any other human being, we necessarily have an idea of who they are and of the relation between us; so do they. When we speak or act towards any other, we do so in terms of our ideas about them and the relation between us; if, for example, we two encounter each other as strangers, then each of us will respond to the other in terms of our specific ideas about strangers and how strangers are meant to comport themselves vis-à-vis others. Our ideas may differ markedly or be fairly similar. Whatever the case, however, I speak to my idea of who you are and you listen and understand me in terms

---

[17] Murillo Pagnotta, pers. comm. (19 Nov. 2015).
[18] Tim Ingold's (2004) arguments for a shift in our understanding of evolution have, likewise, 'centred on processes of development and the dynamic properties of relational fields' (see also Ingold 2007a).

of your idea of who I am. Intersubjectivity necessarily entails that particular, historically constituted ideas of the person are brought into play. Compare the idea of the person as 'an individual' with the idea of the person as a 'locus of relationship' (Toren 2011).

Intersubjectivity is always prior; in other words, it does not arise out of relations between people but precedes them; intersubjectivity is already in play, for example, when a child is born in the sense that the infant's caregivers already have well-formed ideas about what a child is and should be, and about the relation between child and caregiver.[19] The infant's response to the caregiver is assimilated to the caregiver's existing ideas, and in this process transforms them, more or less, depending on the extent to which the infant's responses conform to the caregiver's ideas. The child forms its own ideas of others in the course of its relations with them. The child does not have to be able consciously to represent to itself the other person. Rather, what the newborn infant does is to avail itself of what is made available to it—the mother's nipple in its mouth evokes the sucking response, the softness of her breast and gentle tones of voice are soothing; or it may be the case that she is a reluctant mother, anxious about feeding the child herself, and that the infant feels her tension and discomfort. The infant does not have to represent to itself his or her experiences of being cared for; rather, what happens is that the infant's experience of, say, repeatedly seeing its mother's face and hearing her voice is very rapidly differentiated simply because the infant sees its mother's face over and over again from different angles and with different expressions and under different conditions. The cognitive schema for 'mother' is emphatically not a representation; it is not a definition, nor a kind of mental icon. The schema is best thought of as autopoietic, as a transformational system that is, over time, differentiated in such a way that it assimilates all dimensions of 'mother' that are afforded to the infant—her look, her smell, her feel, her tones of voice, her face, her touch, her sound, the rhythms of her body as she walks, her laugh, and so on, over and over and over again.

In an autopoietic system such as a cognitive scheme, structure and process are aspects of one another, just as they are in any living system. The cognitive scheme is constituted in self-differentiation, which occurs when the scheme is brought into play by the temporal flow of affordances.[20] That is to say, particular affordances are rendered salient by some aspect of what is going on internally or externally to the person. By virtue of their salience in respect of a given person's flow of consciousness, affordances are assimilated to the cognitive schemes in play; this assimilation is at the same time a transformation because the cognitive schemes in play are accommodating to new experiences—always new because always part of the temporal flow of the present. In other words, as Piaget long ago showed, one can think of cognitive schemes as being differentiated in use: whenever a cognitive scheme is in play, it is bound

---

[19] Strictly speaking, one may want to argue that relations between the actual child and the caregiver emerge together (on autonomy and emergence, see Thompson 2007: 59–65). See Pina Cabral (2016) for a fascinating exploration of how naming might figure in the ontogeny of personhood.

[20] My use of 'flow' here comes from Murillo Pagnotta.

to differentiate more or less radically as a function of the relative novelty of what is happening at the time.

Autopoiesis is a process of which it makes no sense to ask whether it is universal because, in every instance, the continuity over time of a species of animal or plant, of a human physical type, or of a schema that structures, say, our perception of faces resides in the continuing differentiation of particular living animals, plants, and human beings. This, I have argued, is inevitably informed by a particular lived history of inter-subjective relations with others in the historically structured environing world in which any given one of us inheres—that is to say, the real world that we each bring into being for ourselves. As I have shown elsewhere, the historical process of human autopoiesis is open to anthropological investigation and analysis.

My proposed unified model of human being is able to come to grips at once with the historical embeddedness of the ideas held by the investigator and by the people who are the object of study because it recognizes that the world provides for all our accounts of it and for their being rendered objective in different ways. It thus makes it easier for the investigator to find out the most basic aspects of people's ontology and entailed epistemology and how the lived world is experienced when these ideas are taken seriously.[21]

I end with a reminder that anthropology is the study of how we humans become who we are. Nothing in this formulation requires an idea of 'human nature' or a distinction between 'biology' and 'culture'. Indeed, it seems to me ever more important that we understand how we come to instantiate our history, for, like all our other ideas, this one has profound political implications for contemporary and future lives. We cannot alter the history that we have lived, but we can perhaps come to a closer understanding of how it is continuing to structure our lives and the lives of our children and grandchildren. It seems obvious to me that an acknowledgement that ontogeny is through and through a historical process would put paid to ideas of 'culture' and 'human nature' and, in so doing, provide for a proper understanding of how certainly we condemn future generations to a deepening of the inequalities and injustices that we are failing to address.

# References

Bamford, R. (2007). *Biology Unmoored: Melanesian Relflections on Life and Biotechnology*. Berkeley: University of California Press.

Biddle, J. (2001). 'Inscribing Identity: Skin as Country in the Central Desert.' In S. Ahmed and J. Stacey (eds), *Thinking through the Skin*, 177–93. London: Routledge.

Blanden, J., and Gibbons, S. (2006). *The Persistence of Poverty across Generations: A View from Two British Cohorts*. Bristol: Policy Press.

---

[21] My unified model of human being addresses Viveiros de Castro's concern that anthropologists 'persist in thinking that in order to explain a non-Western ontology, we must derive it from (or reduce it to) an epistemology' (1998: S79) and Strathern's observations concerning 'the tool science has made of the duplex "relation"...[both positivism and its critiques] put "knowledge" at the forefront of relational endeavour and can imagine different approaches to it' (2005: 42). See also Edwards et al. who bring together ethnographic studies that focus on 'how particular objects of science came onto the agenda as significant, what other ideas they are surrounded by and how ideas travel and have effects in the world' (2007: 10).

Blanden, J., Gregg, P., and Machin, S. (2005). 'Social Mobility in Britain: Low and Falling.' *CentrePiece* 10: 18–20: cep.lse.ac.uk/centrepiece/v10i1/blanden.pdf

Chemero, A. (2009). *Radical Embodied Cognitive Science*. Cambridge, Mass.: MIT Press.

Corak, M. (2013). 'Income Inequality, Quality of Opportunity, and Intergenerational Mobility.' *Journal of Economic Perspectives* 27: 79–102.

Danziger, E. (2010). 'Deixis, Gesture, and Cognition in Spatial Frame of Reference Typology.' *Studies in Language* 34: 167–85.

Danziger, E. (2013). 'Conventional Wisdom: Imagination, Obedience, and Intersubjectivity.' *Language and Communication* 33: 251–62.

Descola, P. (2013). *Beyond Nature and Culture*. Chicago: University of Chicago Press.

Devine, F. (2009). 'Class.' In A. Flinders, A. Gamble, C. Hay, and M. Kenny (eds), *The Oxford Handbook of British Politics*, 609–28. Oxford: Oxford University Press.

Edwards, J., Harvey, P., and Wade, P. (2007). *Anthropology and Science: Epistemologies in Practice*. Oxford: Berg.

Enfield, N. J., and Levinson, S. C. (2006). *Roots of Human Sociality: Culture, Cognition, and Interaction*. Oxford: Berg.

Evans, G. (2006). *Educational Failure and Working Class White Children in Britain*. Basingstoke: Palgrave Macmillan.

Fogel, A. (2011). 'Theoretical and Applied Dynamic Systems Research in Developmental Science.' *Child Development Perspectives* 5: 267–72.

Fogel, A., and DeKoeyer-Laros, I. (2007). 'The Developmental Transition to Secondary Intersubjectivity in the Second Half Year: A Microgenetic Case Study.' *Journal of Developmental Psychology* 2: 63–90.

Fogel, A., Garvey, A., Hsu, H., and West-Stroming, D. (2006). *Change Processes in Relationships: A Relational-Historical Research Approach*. Cambridge: Cambridge University Press.

Franklin, S. (1995). 'Life.' In W. T. Reich (ed.), *The Encyclopedia of Bioethics*, 456–62. New York: Macmillan.

Franklin, S. (2000). 'Life Itself: Global Nature and the Genetic Imaginary.' In S. Franklin, C. Lury, and J. Stacey (eds), *Global Nature, Global Culture*, 188–227. London: Sage.

Franklin, S. (2003). 'Ethical Biocapital: New Strategies of Cell Culture.' In S. Franklin and M. Lock, *Remaking Life and Death*, 97–128. Santa Fe, NM: SAR Press.

Gontier, N., Van Bendegem, J. P., and Aerts, D. (2006). *Evolutionary Epistemology, Language, and Culture: A Non-Adaptationist, Systems Theoretical Approach*. Dordrecht: Springer.

Hess, S. (2009). *Person and Place: Ideas, Ideals, and Practice of Sociality on Vanua Lava, Vanuatu*. New York: Berghahn.

Hollan, D. W., and Throop, C. J. (2011). *The Anthropology of Empathy: Experiencing the Lives of Others in Pacific Societies*. New York: Berghahn.

Husserl, E. (1965). *Phenomenology and the Crisis of Philosophy*. New York: Harper & Row.

Husserl, E. (1970). *The Crisis of European Sciences and Transcendental Phenomenology*. Evanston, Ill.: Northwestern University Press.

Ingold, T. (2004). 'Beyond Biology and Culture: The Meaning of Evolution in a Relational World.' *Social Anthropology* 12: 209–21.

Ingold, T. (2006). 'Against Human Nature.' In N. Gontier, J. P. Van Bendegem, and D. Aerts (eds), *Evolutionary Epistemology, Language and Culture: A Non-Adaptationist, Systems Theoretical Approach*, 259–81. Dordrecht: Springer.

Ingold, T. (2007a). 'The Trouble with Evolutionary Biology.' *Anthropology Today* 23: 13–17.

Ingold, T. (2007b). 'The Social Child.' In A. Fogel, B. King and S. Shanker (eds), *Human Development in the 21st Century: A Dynamic Systems Approach to the Life Sciences*, 112–18. Cambridge: Cambridge University Press.

Leach, J. (2003). *Creative Land: Place and Procreation on the Rai Coast of Papua New Guinea*. New York: Berghahn.

Lepowsky, M. (2011). 'The Boundaries of Personhood, the Problem of Empathy and "the Native's Point of View" in the Outer Islands.' In D. W. Hollan and C. J. Throop (eds), *The Anthropology of Empathy: Experiencing the Lives of Others in Pacific Societies*, 26–43. New York: Berghahn.

Lickliter, R. (2007). 'Developmental Dynamics: The New View from the Life Sciences.' In A. Fogel, B. J. King, and S. G. Shanker (eds), *Human Development in the Twenty-First Century: Visionary Ideas from Systems Scientists*, 11–17. Cambridge: Cambridge University Press.

Lickliter, R. (2011). 'The Integrated Development of Sensory Organization.' *Clinics in Perinatology* 38: 591–602.

Lloyd, G. E. R. (2015). *Analogical Investigations: Historical and Cross-Cultural Perspectives on Human Reasoning*. Cambridge: Cambridge University Press.

Mageo, J. (2011). 'Empathy and "As-If" Attachment in Samoa.' In D. W. Hollan and C. J. Throop (eds), *The Anthropology of Empathy: Experiencing the Lives of Others in Pacific Societies*, 69–94. New York: Berghahn.

Mallett, S. (2003). *Conceiving Cultures: Reproducing People and Places on Nuakata, Papua New Guinea*. Ann Arbor: University of Michigan Press.

Maturana, H. P., and Varela, F. J. (1980). *Autopoiesis and Cognition: The Realisation of the Living*. Dordrecht: Reidel.

Maturana, H. P., and Varela, F. J. (1988). *The Tree of Knowledge*. Boston, Mass.: New Science Library.

Meltzoff, A. N., Kuhl, P. K., Movellan, J., and Sejnowski, T. J. (2009). 'Foundations for a New Science of Learning.' *Science* 325: 284–8.

Miller, P., and Rose, N. (2008). *Governing the Present*. Cambridge: Polity.

Oyama, S. (2000). *Evolution's Eye: A Systems View of the Biology–Culture Divide*. Durham, NC: Duke University Press.

Oyama, S. (2002). 'The Nurturing of Natures.' In A. Grunwald, M. Gutmann, and E. M. Neumann-Held (eds), *On Human Nature: Anthropological, Biological, and Philosophical Foundations*, 163–70. New York: Springer.

Oyama, S. (2010). 'Biologists Behaving Badly: Vitalism and the Language of Language.' *History and Philosophy of the Life Sciences* 32: 401–24.

Oyama, S. (2011). 'Life in Mind: Commentary on Evan Thompson's *Mind in Life*'. *Journal of Consciousness Studies* 18: 83–93.

Pina Cabral, J. (2014). 'World: An Anthropological Examination, Part 1.' *Hau* 4: 49–73.

Pina Cabral, J. (2016). 'Brazilian Serialities: Personhood and Radical Embodied Cognition.' *Current Anthropology* 57: 247–60.

Rabinow, P. (1996). *Making PCR: A Story of Biotechnology*. Chicago: University of Chicago Press.

Robertson, A. F. (1996). 'The Development of Meaning: Ontogeny and Culture.' *Journal of the Royal Anthropological Institute* 2: 591–610.

Rose, N. (2007). *The Politics of Life Itself: Biomedicine, Power, and Subjectivity in the Twenty-First Century*. Princeton, NJ: Princeton University Press.

Stewart, K. (2016). 'The Family and Disadvantage.' In H. Dean and L. Platt (eds), *Social Advantage and Disadvantage*, 85–111. Oxford: Oxford University Press.

Strathern, M. (1981). 'No Nature, No Culture: The Hagen Case'. In C. MacCormack and M. Strathern (eds), *Nature, Culture, and Gender*, 174–222. Cambridge: Cambridge University Press.

Strathern, M. (1992a). *After Nature: English Kinship in the Late Twentieth Century*. Cambridge: Cambridge University Press.

Strathern, M. (1992b). *Reproducing the Future: Essays on Anthropology, Kinship, and the New Reproductive Technologies*. Manchester: Manchester University Press.

Strathern, M. (2005). *Kinship, Law, and the Unexpected: Relatives Are Always a Surprise*. Cambridge: Cambridge University Press.

Thompson, E. (2007). *Mind in Life: Biology, Phenomenology, and the Sciences of Mind*. Cambridge, Mass.: Belknap Press.

Tomasello, M. (2003). *Constructing a Language*. Cambridge, Mass.: Harvard University Press.

Toren, C. (1990). *Making Sense of Hierarchy: Cognition as Social Process in Fiji*. London: Athlone Press.

Toren, C. (1993). 'Making History: The Significance of Childhood Cognition for a Comparative Anthropology of Mind'. *Man* 28: 461–78.

Toren, C. (1998). 'Cannibalism and Compassion: Transformations in Fijian Notions of the Person'. In V. Keck (ed.), *Common Worlds and Single Lives: Constituting Knowledge in Pacific Societies*, 95–115. London: Berg.

Toren, C. (1999a). 'Compassion for One Another: Constituting Kinship as Intentionality in Fiji'. *Journal of the Royal Anthropological Institute* 5: 265–80.

Toren, C. (1999b). *Mind, Materiality, and History: Explorations in Fijian Ethnography*. London: Routledge.

Toren, C. (2002). 'Anthropology as the Whole Science of What It Is to Be Human'. In R. Fox and B. King (eds), *Anthropology beyond Culture*, 105–24. London: Berg.

Toren, C. (2004). 'Becoming a Christian in Fiji: An Ethnographic Study of Ontogeny'. *Journal of the Royal Anthropological Institute* 10: 222–40.

Toren, C. (2005). 'Laughter and Truth in Fiji: What We May Learn from a Joke'. *Oceania* 75: 268–83.

Toren, C. (2007). 'Sunday Lunch in Fiji: Continuity and Transformation in Ideas of the Household'. *American Anthropologist* 109: 285–95.

Toren, C. (2009). 'Intersubjectivity as Epistemology'. *Social Analysis* 53: 130–46.

Toren, C. (2011). 'The Stuff of Imagination: What We Can Learn from Fijian Children's Ideas about Their Lives as Adults'. *Social Analysis* 55: 23–47.

Toren, C. (2012). 'Imagining the World That Warrants Our Imagination: The Revelation of Ontogeny'. *Cambridge Anthropology* 30: 64–79.

Toren, C. (2014). 'What Is a Schema?' *Hau* 4: 401–9.

Tui Atua, T. T. T. E. (2009). 'Bioethics and the Samoan Indigenous Reference'. *International Social Science Journal* 60: 115–24.

Viveiros de Castro, E. (1998). 'Cosmological Deixis and Amerindian Perspectivism'. *JRAI* 4: 469–88.

Vygotsky, L. S. (1986). *Thought and Language*. Cambridge, Mass.: Harvard University Press.

Walton A. E., Richardson M. J., Langland-Hassan P., and Chemero A. (2015). 'Improvisation and the Self-Organization of Multiple Musical Bodies'. *Frontiers in Psychology* 6: https://www.frontiersin.org/articles/10.3389/fpsyg.2015.00313/full

# 10

# Divide and Conquer
The Authority of Nature and Why We Disagree about Human Nature

*Maria Kronfeldner*

## 10.1 Seven Reasons Why We Disagree about Human Nature

There are at least seven reasons why we disagree about human nature.* The first two are connected to what I call the politics of human nature, and the rest are connected to scientific issues. The last of these reasons is the focus of this chapter.

First, human nature is about 'our' nature. David Hull (1986: 6) noted that we often describe other species in a careful statistical and non-normative manner, but when it comes to our species, we often fall back into essentialist traps, involving normalcy and normativity. Hull regarded this 'coincidence [as] highly suspicious'. In the words of Proctor (2003: 220), we do not ask about an entity 'being "fully cockroach" or "fully chimpanzee"', but we do regard some humans as more fully human than others, or as realizing more natural goodness. The source of this exceptionalist way of dealing with our nature lies, first and foremost, not in any epistemic functions of the concept, but rather in its normative function for us, which is after all a political function. With respect to this function, the concept is essentially contested in the sense of Gallie (1956): the only essence in that concept is that it is

---

* I want to thank the Sydney Centre for the Foundations Science, the Fishbein Center for History of Science and Medicine at the University of Chicago, the Department of Philosophy at Bielefeld University, and the DFG-Netzwerk Philosophie der Lebenswissenschaften (KR 3392/2-1) for supporting and funding the research for this chapter. Thanks to the audiences at these institutions, to my students of the 2015 Human Nature course at the Central European University, and to the support from the ERC project 'A Science of Human Nature?'. Special thanks go to Martin Carrier for his generosity in supporting my research. I also want to thank Stephen Gaukroger, Jackie Feke, Paul Griffiths, Evelyn Fox Keller, Tim Lewens, Geoffrey Lloyd, Diane Paul, Michaela Rehm, Alexander Reutlinger, Robert J. Richards, Neil Roughley, Karola Stotz, Georg Toepfer, and Anik Waldow. They all contributed in one way or another to the genesis of this chapter. Thanks to Elizabeth Hannon for very helpful editing.

contested. In terms of a slogan: by continuously contesting what it means to be human, we continuously become human.[1]

Second, the history of the vernacular (or folk) concept of human nature suggests that 'being human' is an empty category that simply says, in the words of Marshall Sahlins (2008), 'L'espèce, c'est moi.' If 'human nature', in a broad descriptive sense, simply refers to 'what it means to be human', then this concept has been used—historically and in different cultures—for whatever characterizes the respective in-group. The respective out-groups are consequently dehumanized, that is, regarded as less human.[2] Evidence from historical, anthropological, and psychological scholarship supports this claim about the exchangeability of the content (or 'indexicality', as Smith 2013 termed it). As a result, the content of the concepts 'human' and 'human nature' varies, even when the political function of social demarcation—of demarcating the in-group from the out-group—stays the same. Consequently, different human groups will (explicitly or implicitly) disagree about 'what it means to be human' and thus about human nature.

Contemporary scientific approaches that use the concept of human nature will try to prevent the exchangeability of the content of the vernacular concept.[3] Nonetheless, scientific approaches will face their own set of disagreements, based on differences in usage. Usage differs, for instance, depending on the epistemic goal, since from the scientific point of view, the term 'human nature' can be used for the purposes of description, explanation, or classification. In turn, these goals guide the production of knowledge. The term 'human nature' can, accordingly, refer to one of three 'natures':

- a descriptive nature: a bundle of properties describing the respective group's life form, that is, what it means to be human;
- an explanatory nature: a set of factors with explanatory relevance for the respective life form;
- a classificatory nature: classificatory criteria that determine the boundaries of, and membership in, a biological or social group called 'human'.

---

[1] For human nature as an essentially contested concept, see Kronfeldner (forthcoming). 'Becoming human' by contesting the meaning of human nature involves not only 'making meaning' (e.g. in the sense of Toren, Ch. 9 this volume) but also 'making people' (in the sense of Hacking's (1995) looping effects).

[2] See Kronfeldner (2016), where I discuss the connection between dehumanization and human nature in detail. Dehumanization does not require a concept of human nature in a narrow sense (i.e. contrasted to culture); all it requires is either a graded genealogical association (people as more or less genealogically related) or a differential attribution of properties deemed to be central for 'what it means to be human'. However, the idea that these properties are part of a human nature is a catalyser for dehumanization.

[3] With this and the following, I restrict my analysis to the history of the tradition from which modern science emerged. There are ways of setting things up differently, however, e.g. on the basis of alternative ontologies, such as those described by Descola (2005). For a comparison and discussion of traditions other than those of the West, see Lloyd (2012, 2015). If these traditions are taken into account, at least two further reasons for disagreement come to the fore: different ontologies (process ontology, relational ontology, substance ontology, and so on), as addressed by Dupré (Ch. 5 this volume) and Toren (Ch. 9), and different meanings of the nature/culture divide. Space does not suffice to discuss these issues.

According to the traditional essentialist picture, an essence or nature is, by contrast, a thing that fulfils all three of these epistemic roles simultaneously. An essence is first and foremost classificatory and explanatory. It is what 'makes' individuals human, definitionally and causally. The descriptive role is derivative but covered too, since the description of the properties that are characteristic of the respective kind at issue (e.g. human beings) is the explanandum, i.e. what is explained by the essence. Thus, if you do not yet know the characteristic properties of the kind and you learn about the 'essence', then you can derive the properties explained by the essence. In addition, 'essence' often had a normative connotation: what is part of human nature is not just classificatory and explanatory (and derivatively descriptive), but also what a 'normal' human should exhibit.[4]

In the current philosophy of science literature on human nature, there is only one significant consensus, namely, that traditional essentialism is wrong. What remains is a pluralism of human nature concepts.[5] This pluralism, derived from anti-essentialism, has a couple of important aspects that can help us to understand why, even within science, we can (and likely often will) disagree about human nature.

First, there is a plurality of referents for the term 'human'; often the assumption is that it refers to members of the species *Homo sapiens*, but even that is contested.[6] Consequently, the term 'human' can be interpreted to refer to recent humans only, to a larger biological group (e.g. to one including the Neanderthals), or even to a purely social group (e.g. to all those able to communicate and interact in rational and moral ways with others). None of these interpretations is, in and of itself, more scientific or objective than any other, and which referent is chosen depends on the disciplinary focus. Sociology is unlikely to be interested in biological groups and will focus on social groups. Biologically oriented disciplines will tend to focus on one of the biological groupings. This 'relativity of human nature', as Machery (Chapter 1 this volume) calls it, causes a considerable amount of disagreement. Without fixing the group to which the term refers, everything else about 'human nature' will float around loosely. Unfortunately, in many discussions the referent is left implicit.

Second, there is disagreement about the classificatory criteria, even if there is agreement on the reference to a respective group. Hull (1986), for instance, takes the genealogical nexus (i.e. the genealogical relations between people) as the classificatory

---

[4] The connection between the normative role and the three epistemic roles is not easy to capture, especially since it has a long history, with all the variation that comes with that. As Lloyd makes clear: 'When certain phenomena or practices are labelled "unnatural", that is sometimes just an expression of disapproval [...] with no reference to how frequent ("normal" in that sense) the "unnatural" may be. The antecedents of that use go back (again) to Aristotle [...] What is *para phusin* can be more common than what is *kata phusin*' (pers. comm., 25 Nov. 2015).

For general discussion of the normative role of human nature and how it connects to the concept of 'normal', see Foot (2001), Thompson (2008). For a critical take on it, see Antony (1998, 2000); Silvers (1998).

[5] As described in detail in Kronfeldner et al. (2014). Here I will only summarize the resulting pluralism and add the points that relate to the kinds of disagreements resulting from it.

[6] See e.g. Smith (2013).

criterion for delineating *Homo sapiens*; others disagree, opting for a cluster of properties to delineate the species. By contrast, if 'human' refers to a social group, a social nexus (analogous to the genealogical nexus) or social trait clusters can be taken as definitional.

Third, people are unlikely to fully agree on which properties are part of the descriptive nature. Darwinian ontology tells us that variation is not just ubiquitous in all biological species, but necessary for evolution to occur. Thus none of the traditional, intrinsic candidate properties for a descriptive human nature—rationality, intentionality, morality, language, and so on—are strictly instantiated by all and only humans. Furthermore, depending on what evidence is taken into account, any claim about a property's typicality and uniqueness can be challenged. One researcher might stress that non-human animals are also rational, though not moral; another might argue that they are moral in some sense, but do not have the same intentionality. Can there ever be an end to this kind of disagreement, given that it is likely that whatever property we choose, we will eventually find something similar in other animals if we search for it hard enough? Furthermore, since these candidate properties are presumably all connected, and since no one property describes better than any other what is typical and/or unique about the human life form, there is a choice involved if people focus on specific candidate properties that 'make us human'. Taking these two problems together (one about evidence, one about relationships among the properties), an interesting underdetermination results: certain properties can be prioritized without science providing any objective foundation for this priority. Some researchers will highlight rationality, others morality, still others the opposable thumb, and so on. As is often the case with underdetermination, both disciplinary focus and social values affect what is considered to be the most important property (or properties) of 'being human'; and there is often no way to find agreement on this choice from within science.

Fourth, a descriptive property (or property cluster) is not necessarily the same thing as a classificatory criterion. For instance, relations between people in a group are simply not the same as the cluster of properties characteristic of that group. Thus, if the classificatory criterion is the genealogical nexus, then that 'nature' is simply not the same as the descriptive nature. Equally, clustered properties are not the same as the factors explaining these properties. Consequently, there are different things in the world that we can call a 'human nature'. It follows that the term 'human nature' can refer to whatever it is that fulfils the classificatory role, the descriptive role, or the explanatory role. As Kronfeldner et al. (2014) show, none of the candidates for a post-essentialist nature can do the same epistemological work as traditional essentialist accounts, since none of them will fulfil a classificatory, descriptive, and explanatory role simultaneously. Disagreement among the anti-essentialists (i.e. those that agree that traditional essentialism does not apply to biological species) results mainly from this post-essentialist pluralism.

Fifth, even though the disagreement rests on the anti-essentialism of the field, it is often catalysed by the normative dimension of the term 'nature' in the sciences.

At the core of this normativity is the idea of 'nature' as a contrastive term. While some prioritize what they call 'human nature', others prioritize its contrast—for instance, human culture. Since scientists disagree on what is important, given the disciplinary structure of science (among other things), they also disagree about the importance of human nature as an explanatory principle, despite the fact that their approaches can be understood as complementary. Some want to appropriate the term for what they consider to be important, while others oppose it because they deem something else to be important. For instance, 'nature' in the explanatory sense will be important for explanatory fields such as cognitive psychology; 'nature' in the descriptive sense will be important for fields that simply want to describe humans (e.g. anatomy, physiology); others still (e.g. cultural anthropologists) will regard 'nature' as explanatorily and descriptively unimportant for what they study. In the rest of this chapter, I will show in detail that there is an inherited, normativity-generating authority in the term 'nature' that derives from the pragmatics of legitimizing one's style of inquiry in the marketplace of knowledge production. The importance of appropriating the term 'nature' for the causes and phenomena one is studying is analogous to the importance of appropriating the term 'truth' for the resulting claims. To call what one studies 'nature' is to give one's research a seal of quality and signal its importance, just as calling one's research findings a 'truth' has traditionally indicated its quality.

It is because of this pragmatic dimension that different parties appropriate the term or oppose its use; it is used to highlight their research and what it is they believe to be important. Since there will be no agreement on what is scientifically important (as this is usually determined by the scientist's particular research interests), there will be no agreement on what constitutes human nature, despite agreement on matters of fact.

I will argue that this fifth scientific source of disagreement, resulting from the authority that the term 'nature' imbues, is the reason why some contemporary post-essentialist accounts—even those in this volume—are unlikely to agree on how we should use the term 'human nature'.

## 10.2 The Importance of Understanding the Authority Inherent in the Term 'Nature'

If the pragmatic function of a concept is ignored, it will be hard to see why people care, why people fight for or against a concept or specific term. To understand the pragmatic function, we need to understand the authority that the term 'nature' imbues. The claim that I will defend in the following is an extension of what Geoffrey Lloyd (1991: 432) claimed for the concept of nature in Greek antiquity:

the idea of nature was supposed to stand simply for what is there, for what can be taken for granted. Yet what that comprised was repeatedly contested, not just so far as the natural world in general went, but also as far as human nature is concerned [...] Nature was what was presupposed to be there to investigate: its supposed objective reality was what guaranteed the viability of the investigation. Yet what that vaunted objective reality consisted in was contested in every conceivable respect.

## 10.3 Nature?

Even though there is a mind-boggling variety of meanings attached to the term 'nature' (derived from the Latin *natura*, and going back to the Greek *physis*), two aspects are quite basic according to Lloyd (1991): that 'nature' refers to the nature of things, and that it refers to the things of nature. The first meaning concerns the essences of kinds, which has strong connections to growth and reproduction, etymologically. The things of nature, by contrast, can be understood as those things that can be investigated in a systematic manner, oriented towards accessible evidence. The rainbow and further 'natural' phenomena were the first things in the Western history of philosophy that became naturalized in this sense. The moment they were no longer conceived of as mythological, they became part of nature.[7]

What unites these two basic meanings is that since Greek antiquity they have both been used in a dualistic (i.e. antithetical) manner. They carry a contrast: natural versus supranatural and natural versus cultural, to name just two of the contrasts that form part of a dualistic landscape.[8] The contrast between nature and culture is under much attack, including in this volume. Elsewhere, I defend the contrast against its critics (Kronfeldner forthcoming), but here I aim to analyse (rather than criticize or defend) the nature/culture divide as one instantiation of the contrastiveness of the term 'nature'.

## 10.4 A Core Claim and Two Main Follow-Up Theses

There are pragmatic reasons for these contrasts: they are used not only to demarcate phenomena and to explain those phenomena, but also to demarcate expertise over the phenomena and explanations, i.e. to establish epistemic authority for a special group of people.

I will establish this core claim by looking at historical cases from Greek antiquity, Enlightenment philosophy, the advent of the study of heredity at the beginning of

---

[7] The term 'nature' is quite fundamental in many diverse philosophical fields. That might explain why it has so many meanings and why, interestingly, there are no articles on 'nature' in any of the major philosophy encyclopedias (e.g. the *Stanford Encyclopedia of Philosophy* or the *Routledge Encyclopedia of Philosophy*); 'nature' might be too fundamental. For an insightful and classic discussion, see Mill (1874), as well as Collingwood (1945) and Lewis (1960).

[8] In feminist and post-structuralist discourse, the contrastiveness of nature is often regarded as hierarchical: '"Otherness" entails boundaries, exclusions and inclusions policed by categories and rules [...] Otherness is not reciprocal' (Haste 2000: 177). This means that nature is primary and that the contrasts depend on it. What I defend here is compatible with such a claim about hierarchy, but does not rely on it.

the twentieth century, and contemporary evolutionary psychology. Each case highlights a time when the term 'nature' was used in contrast to something else and was used pragmatically, to assert authority.

The first follow-up thesis that can be derived from looking at the historical cases is that even if the assumed contrasts to, and the contents of, talk about 'nature' (in the context of understanding human life) varied historically, the pragmatic function of demarcation and exclusion—to demarcate and then exclude kinds of explanations and the experts offering them—stayed the same, at least in the cases mentioned. The second follow-up thesis will be that given this history, it is plausible that the term 'nature' is often used in this pragmatic sense, namely, to gain authority for one's style of inquiry—i.e. the methods one uses, which also depend on what one regards as important. Since styles of inquiry will vary, the term's referent will vary too, and there will be no way to settle on one variant.

## 10.5  The Invention of Nature in Greek Antiquity

> ...some of those who insisted on the category of the natural [in Greek antiquity] used it to demarcate and justify their style of inquiry, their methodology, in contrast to those of rivals whom they were hoping to put out of business.
>
> (Lloyd 1991: 422)

In this quotation, Lloyd is referring to some of the Hippocratic authors, and the contrast at issue was that between the natural and the supranatural. The author of the Hippocratic treatise *On the Sacred Disease*, for instance, claimed that all diseases are naturally (rather than divinely) caused, including the 'sacred disease', a disease probably most similar to what is now called epilepsy. It is a disease, as the author of the Hippocratic treatise claimed, that is 'nowise more divine than others [...] it has its nature such as other diseases have, and a cause whence it originates, and its nature and cause are divine only just as much as all others are, and it is curable no less than the others' (Hippocrates 1849: 847).

At that time, such all-inclusive claims about natures and causes tended to be made on the basis of faith, rather than because of any knowledge of facts, even though the Hippocratic authors certainly had some knowledge about diseases, and offered sketches of naturalistic explanations for particular diseases. They used these sketches (and the related all-inclusive claims) to argue against those whom they wanted to put out of business—that is, those who believed that the cause of disease was divine visitation and that the cure lay in ritual purification and incantation. There was often no consensus about what knowledge was available, even among the Hippocratic authors and practitioners. Thus, since the positive explanations were sketchy, often rather speculative, and controversial, the naturalness of all diseases was assumed. This was important for the demarcation of their business and the exclusion of others from it. The Hippocratics not only offered arguments against those attributing the disease to divine influences, the 'con-jurors, purificators, mountebanks, and charlatans' (Hippocrates 1849: 841), but also complained that their enemies were in the business of understanding diseases for money. However, the Hippocratics were in it for money as well, and sometimes for a lot (Lloyd 1978: 18–19). In the words of the Hippocratic treatise *The Law*, the practical problem was that 'in the cities there is no punishment connected with the practice of medicine' (Hippocrates 1849: 784). By contrast, the Babylonians had the Hammurabic Code, and Egypt had a state-controlled medical hierarchy (Porter 1997: 53f.). These institutions regulated and, in so doing, demarcated the behaviour of those in the business of healing people. Given that in Ancient Greece there was no legal means to demarcate the Hippocratic style of inquiry from temple medicine—for instance, by academic titles or similar qualifications—the battle of demarcation had to be fought by different means.

The tools that were used as a stand-in for legal methods were conceptual. The Hippocratics utilized their epistemic tools to defend their lot from their competitors. They established an explicit and general concept of nature that they called *physis*. This was a concept that Aristotle (influenced by the Hippocratics) brought to full bloom, but that earlier Greek writers—Thales, Anaximander, Anaximenes, Xenophanes, and so on—were already using. An important point to keep in mind, however, is that the way we interpret these philosophers now tends to be anachronistic: it is only given later developments that pre-Socratic philosophers appear as natural philosophers, as we now call them. If we try to keep this in mind, Lloyd (1991: 418–20) claims, then what was new in the Hippocratic treatises and then in Aristotle was that *physis* had changed from merely referring to the nature of things to also explicitly referring to the things of nature—that is, to a general 'domain of nature' (p. 420). Lloyd admits that there had been some idea of a domain with 'regularities of what *we* call natural phenomena' (p. 419; emphasis added) already in existence, but 'there is all the difference in the world between an implicit assumption and the *explicit* concept' (p. 419; emphasis original).[9]

Consequently, that the Greeks (and not the Babylonians nor the Egyptians) developed an explicit, general, and systematic concept of naturally caused versus supranaturally caused things is less surprising if we take into account that they could make practical, pragmatic use of it: they could use it for social demarcation and exclusion. There are, of course, other factors relevant in the explanation of why the general notion of nature was invented in the context of Greek antiquity and not elsewhere. Nonetheless, epistemic and social demarcation was an important function, it seems. In Lloyd's (1991: 432) words, 'the category of the natural was used to legitimate a point of view specific to the interests of some particular group'. Hippocratic authors organized around naturalism

---

[9] For further details and justification, see also Hager's section (1971–2007) on Greek antiquity in the entry on 'nature' in *Historisches Wörterbuch der Philosophie*. Lloyd's claim is compatible with Lehoux's (2012: 57), namely, that the idea of laws of nature (even in its non-modern form) was not yet part of the Greek 'invention of nature'. For more on the connection between Greek philosophy and Hippocratic medicine, see Lloyd (2003).

and against any divine interference, uniting—despite disagreement about the precise causes of particular diseases—against an enemy (theurgy, magic) in the marketplace of healing people.

The content of the concept of nature at this time, in the context of explaining the traits of people (e.g. the 'sacred disease'), depended on its contrast with the supernatural. The pragmatic function of the concept and of the term was to legitimize styles of explanation and to appropriate phenomena—to define and circumscribe a specific medical practice. In this way, the concept (and the term) 'nature' *divided* (kinds of causes, methods to study these causes) and thereby *conquered* (competitors for authority over certain phenomena). Dividing kinds of practices entails dividing kinds of causes (in this case, natural and supernatural). Thus the pragmatic function of both concept and term was connected with a specific epistemic role of the concept, namely, explanation: via demarcation, it was decided which kinds of causes were relevant for the explanandum.

The pragmatic function of demarcation is thus not only social, but epistemic. This is so in two senses. First, it is concepts (rather than titles or degrees) that are used for demarcation. Second, the demarcation does something to our epistemic activities (such as causal explanations): it legitimizes them as adequate. The concept of nature thus acted as a framework, defining kinds of causes via contrast (natural versus supernatural) and deciding which causes were relevant and which tools (medical practice, Hippocratic style of inquiry) could and should be used to study the phenomena at issue (in this case, diseases).

Why is this relevant for a discussion of human nature? The term 'human nature' is seldom used these days with the contrast between natural and supernatural in mind. The concept has changed. However, the Hippocratics' use of it in Greek antiquity gives support to the idea that 'nature' can be used to define kinds of relevant causes and thus function as an authority-granting epistemic tool, determining who can legitimately study and claim expertise about a given phenomenon. In the following, I will provide further evidence for this idea.

## 10.6 Enlightenment Philosophers and the Science of Human Nature

Roger Smith (1995, 1997) found a similar process of demarcation and exclusion when human nature became a major philosophical topic during the Enlightenment.[10] Again, it was a common enemy that united those involved. For the Enlightenment philosophers interested in a naturalized and empirical 'science of man', the enemy was metaphysics in general, and Christianity and Renaissance humanism more specifically.

As with the Hippocratics, the concept of nature was treated as an a priori concept: that there is such a nature was taken for granted. Furthermore, the concept defined

---

[10] Smith writes: 'To quote references to human nature in the eighteenth century is a bit like quoting references to God in the Bible: it is the subject around which everything else revolves' (1997: 216).

an area of study, the new 'science of man'. The concept of nature, and in this case 'human nature', once again determined the framework within which diverse content or explanations were discussed and appropriated. The dominant contrast was between empirical investigation and speculative metaphysics. A science of human nature did not necessarily entail a naturalization of human nature or thinking about it in physicalist terms, but the new experimental methods stimulated interested philosophers to seek out systematic, empirical knowledge about humans.

David Hume, for instance, used the term 'human nature' to specify the target of the envisioned 'science of man' and, within that framework, aspired to do for moral philosophy (the 'science of man') what Bacon, Galileo, Newton, and the like did for natural philosophy (Smith 1995, 1997). Even though Hume hoped to develop an 'accurate anatomy of human nature' (Hume 1978: 263), his aim was not to equate human nature with physical nature. The general demarcation at issue was between the practice of a science of human nature and metaphysical speculation. The contrast (and thus the content of the concept of human nature) was not between the natural and the supranatural, but between empirical investigations and speculative metaphysics. Yet the pragmatic function stayed the same: the demarcation and exclusion of styles of inquiry.

The term 'nature' in 'human nature' had two different epistemic roles to play in the context that Hume was invoking, in my terms, both a descriptive and an explanatory nature. As a descriptive category, the term was used, as just mentioned, to describe the explanandum of the 'science of man'. Hume was looking to establish a science with human nature as the explanandum, a science that would uncover our typical and species-specific ways of being (the human life form), which in turn would be subject to law-like generalizations similar to the laws of nature. Simultaneously, the term 'nature' was used as an explanatory category. It referred to capacities, and the goal was to explain overt behaviour in terms of a nature (a set of capacities) that was atemporally given, and more or less inhering in (i.e. internal to) more or less all people. In his famous account of causation, for instance, Hume argued that our causal inferences rely on habit, and that habit was part of human nature (i.e. how humans naturally function). Thus Hume held that if we want to understand why we talk and think about causation the way we do, then we have to understand human nature. However, internal natural capacities such as habit were not themselves Hume's target of explanation; they were taken as generalizations about humans that were assumed to be part of the explanans, i.e. part of what explains human behaviour (i.e. causal reasoning). He used the term 'human nature' to refer to an explanatory nature that was assumed. According to Hume, we cannot study this explanatory nature, since hypotheses about its cause should 'be rejected as presumptions and chimerical' (Hume 1978: xvii). Thus habit is 'a principle of human nature, which is universally acknowledged, and which is well known by its effects' (Hume 1975: 43); it is a 'primitive element' of human understanding, as Norton (1993: 158) writes, interpreting Hume as I do here.

Smith (1995, 1997) concludes from Hume and other Enlightenment philosophers' talk about human nature that a shared aim at that time was to describe the characteristics of humans in an empirical—i.e. in an evidence-based and systematic manner—rather than in a speculative, metaphysical manner. Thus the same kind of demarcation and exclusion as in Ancient Greek medicine is intended, even though the contrast has changed somewhat.

But how was it the case that 'human nature' was able to play this demarcation function? First, this was all that these Enlightenment thinkers could agree on, since opinion on the precise contents of what it 'means to be human' varied widely. Hobbes rallied for the egoistic man, Rousseau for the noble savage, and La Mettrie for man as a machine, to name just three philosophers involved in Enlightenment discussions about human nature. Smith (1995, 1997) concludes that during the Enlightenment, the concept of a 'human nature' helped to create and maintain a common language, a framework for discourse, which in turn allowed for discussion of the differences in opinion about the specific qualities of human nature (as explanandum). It helped to unite and demarcate a specific kind of Enlightenment philosophy, just as the concept of natural diseases helped the Hippocratics to unite and demarcate. Second, modern science—an institution of systematic and empirical study—was being established at this time, but did not yet have a secure enough foundation from which to defend its style of inquiry against speculative philosophy. As with the Hippocratics, in the absence of secure legal or other institutional means with which to defend one's epistemic authority over a subject matter, concepts—epistemic tools—were used to demarcate and exclude styles of inquiry.

The historical conclusion that I want to draw from this is that the content of the concept of human nature may have varied, but the pragmatic function remained the same: the demarcation and legitimization of a specific style of inquiry, and the exclusion of other styles as irrelevant or illegitimate. Having used only two cases, the conclusion I can draw is limited. But the comparison of the two cases gives us something at least: the contents of the concept of 'nature' or 'the natural' varied, whereas the demarcation function of the concept stayed the same. The concept (and term) 'nature' united people against an enemy with a different style of inquiry. And in both cases, this demarcation was connected to a distinction between different kinds of explanations. Even so, the contents only varied narrowly, since the contrast at issue was similar. I will now turn to a clearly different contrast, that between *physis* and *nomos*, or nature and culture.

## 10.7 *Physis* and *Nomos* in Greek Antiquity

As with the origin of the contrast between natural and supranatural, the origin of the contrast between *physis* (nature) and *nomos* (culture) is believed to lie in Greek antiquity. Has the distinction also been motivated by the need for demarcation between styles of inquiry? The author(s) of *Airs, Waters, and Places* might have believed something like, 'we medical scientists will take care of *physis*, physical characteristics, while you philosophers can take care of *nomos*'. Unlike the divide between the natural and supranatural, finding evidence that the divide between nature and culture had a social demarcation role in Greek antiquity is difficult.

In Greek antiquity, the contrast between *physis* and *nomos* was not yet fully developed. First and foremost, the Hippocratic authors—for example, those of *Airs, Waters, and Places*, as well as those of *The Sacred Disease*—assumed an inheritance of acquired characteristics. Indeed, the inheritance of acquired characteristics was assumed by almost everyone in Greek antiquity; it was simply common sense, and stayed so well into the nineteenth century.[11] If acquired characters can be inherited—if *nomos* successively turns into *physis*—then *physis* and *nomos* are so entangled that demarcation based on them seems unlikely.

Second, philosophers had not fully developed a strong contrast between these two ideas either, primarily because of the teleological way nature was conceived of at that time—a point that is of utmost importance for understanding the historical connections between nature, natural law, and laws of nature.[12] Nature (often capitalized, to refer to the 'things of Nature') and the natures of things were connected to a goal or *telos*. For Aristotle, the connection between *physis* and *nomos* was thus quite tight. In his *Politics*, Aristotle spells out that by nature, humans form male–female bonds, then households, and then, to secure the reproductive bonds and those things necessary to make that bond successful, they form the Greek *polis*, with its laws and customs—that is, *nomos*. It is a part of fulfilling our nature (i.e. our *physis* and natural *telos*) for humans to form male–female bonds, households, a *polis*, and *nomos*. Thus it is not only the case that Aristotle can be understood as the first theoretical biologist (his biological treatises often contained more scientific detail than the Hippocratic treatises); he also regarded *nomos* (part of the life form and *telos* of the human species) as an integral part of nature, not as contrasted with it.

Certainly, Aristotle is just one example, and it is almost sinful to ignore all the other philosophers of his time who also had something to say on the matter of *physis* and *nomos*[13]; but given what we know about the situation generally, it would be a surprise if the dichotomy were fully developed at that time. In addition to the fact that the belief in inheritance of acquired characteristics was standard, there was no need to use the divide in order to exclude a certain group of people from a certain epistemic practice (be it descriptive, classificatory, or explanatory), since biology and political philosophy were mostly the preserve of the same set of people, namely, Greek philosophers like Aristotle. Granted, there were the Hippocratics, specializing in diseases; but (to the best of my knowledge) the philosophers seem to have had no stake in, and didn't interfere with, the business of these experts. In turn,

---

[11] See Zirkle (1946) on the long history of the belief in inheritance of acquired characteristics.
[12] See Collingwood (1945) on nature teleologically conceived.
[13] See Heinimann (1945) for a book length in-depth analysis of the contrast in Greek antiquity.

these experts had other enemies to fight in the marketplace of healing people, as illustrated above.

Things changed towards the end of the nineteenth century. The divide between nature and culture hardened, and became generalized as 'nature versus nurture'. This development created a gap or, as in the title of Evelyn Fox Keller's (2010) book on this subject, a 'mirage of space between nature and nurture'. Increasingly, this gap was perceived as unbridgeable. The hardening of the divide was again connected with a marketplace, no longer the *polis* but the developing institutions of academia, with its emerging disciplinary separations. The divide between nature and culture came to mark the boundaries of different academic fields, helping to make disciplines and the divisions of authority clear-cut.

## 10.8 The Advent of Heredity and the Rally against Lamarckian Inheritance

According to the received view, Charles Darwin's cousin Francis Galton introduced the 'modern' nature/nurture divide (a variant of the nature/culture divide). He famously used the phrase 'nature and nurture' as 'a convenient jingle of words, for it separates under two distinct heads the innumerable elements of which personality is composed. Nature is all that a man brings with himself into the world; nurture is every influence from without that affects him after his birth' (Galton 1874: 12). Galton believed that the 'distinction is clear' (p. 12). In the context in which he was writing, the distinction was indeed clear, since it was a context that had heredity as a new field-structuring explanandum, uniting people against its opponents. And, as before, words (Galton's 'convenient jingle of words') were important to mark the boundary. The term 'nature' (here referring to hereditary developmental resources) and its contrast (nurture as an inclusive term for culture, environment, and everything else not transmitted via biological reproduction) became crucial to defending the line between those studying biological heredity in a new (i.e. statistical as well as experimental) manner and those doing something else.

In the nineteenth century, historians such as López-Beltrán (1994) claim, there was an intellectual shift: reference to the adjective 'hereditary' (as in 'hereditary disease') was increasingly replaced by a nominal use of the noun 'heredity' as a field-defining phenomenon in need of explanation. This amounts to a reification, bringing with it new ontological commitments and the creation of, in Müller-Wille and Rheinberger's (2007) terms, a new 'epistemic space'. But most importantly,

[it] also implies a concomitant shift, namely the erosion of a set of very ancient distinctions with respect to similarities between parents and offspring, which the modern notion of heredity systematically cuts across. Distinctions had been made between specific versus individual, paternal versus maternal, ancestral versus parental, normal versus pathological similarities,

and even between similarities pertaining to the left and the right halves of the body. Such distinctions gave way to a *generalized notion of heredity* that focused on elementary traits or dispositions independent of the particular life forms they were part of, whether pathological or normal, maternal or paternal, individual or specific.

(Müller-Wille and Rheinberger 2007: 13; emphasis added)

Galton replaced these older distinctions with a new generalized distinction, the nature/nurture divide. The establishment of this contrast was strongly influenced by Galton's anti-Lamarckism.

Galton developed the idea of particulate inheritance—an idea already discussed by Charles Darwin—which took biologically inherited developmental resources as material substances—'gemmules' in Darwin's case and 'stirps' in Galton's. For Galton, the hereditary units were material and internal to individuals, as they were for Darwin; *contra* Darwin, however, for Galton they were also fixed, unchangeable—as ahistoric as the units of the physical and chemical world. They are the 'elements of which personality is composed' (Galton 1874: 12). As elements, they cannot be changed during individual development. Thus, inheritance was 'hard' (relying on unchangeable elements), rather than 'soft' (as in the Lamarckian picture). Consequently, Galton's distinction between stirp and person (between latent and patent elements) counts as predecessor of the germ/soma distinction, later introduced by Weismann. With respect to Lamarckian inheritance, Galton stood against most of his peers, and also against his cousin Charles Darwin. Others, such as Wallace and most famously Weismann, later joined in to form a front against Lamarckian inheritance.

Using the term 'nature', with its history of playing an authority-establishing demarcation role, was part and parcel of Galton's theory of particulate and hard heredity. Since Galton's view on heredity and the nature/nurture contrast also involved a specific style of inquiry (namely, statistics and the use of twin studies, combined with experimental studies), we can regard the case as analogous to the other case studies: 'nature' is used to demarcate, though the demarcation does not make distinct disciplines so much as distinct styles of inquiry in the emerging study of heredity, development, and evolution. In conclusion, this nineteenth-century case tells a similar story to those of the previous two.

## 10.9 The Formation of Cultural Anthropology as an Academic Discipline

Mirroring the formation of heredity as a field-defining explanandum, 'culture' played a similar role, and did so at roughly the same time (the turn of the twentieth century), when cultural anthropology began to assert its identity among the other aspiring scientific disciplines (mainly the new experimental genetics and psychology), as well as the traditional physical anthropology, and a tradition of museum-based anthropology with its strong racist leanings. In the mid-nineteenth century, anthropology was not yet a separate academic discipline, even though interest in studying the differences between cultures dates back to at least Greek antiquity; it arises, for example, in the Hippocratic Treatise on *Airs, Waters, and Places*.

Edward B. Tylor (1871), often treated as the grandfather of scientific anthropology, saw 'culture' contrasted with 'nature', but culture was mainly an explanandum. In other words, the concept of culture in the mid-nineteenth century did not yet serve the function of demarcating causes: it was not used to divide kinds of causes to explain behaviour, either explicitly or at the theoretical level. There was an awareness of different kinds of causes influencing development, but the dominant pragmatic function of the concept of culture was to define kinds of people (rather than kinds of causes), and to point at historical change (often called 'civilization') or the emergence of institutions. Culture described behaviour (the explanandum), but it was not yet used as the label for a field-defining style of explanation that contrasted nature with culture. And how could it be otherwise, given that Tylor, as most people at his time, believed in the intricate connection between nature and nurture through the inheritance of acquired characteristics?

It was Alfred L. Kroeber who radicalized the nature/culture divide. He took the perspective of his teacher Franz Boas, who studied, among other things, the influence of culture on physical traits among immigrants to the United States, as illustrated by Stocking (1968: 195–233). Boas explicitly treated culture as a specific explanatory factor in the development of the traits of individuals. Kroeber went further, taking culture to be not only a factor in the development of individuals, but a field-defining explanandum in its own right. As with Tylor, culture was the explanandum, but one that was explicitly, theoretically, and completely decoupled from nature. This was the basis of his 'cultural determinism': culture is explained by culture alone, and what is inherited by nature is explained by nature alone.

Kroeber felt the need to demarcate his own and his peers' business from that of others. Kroeber and his fellow anthropologists had degrees to secure their intellectual authority (in fact, he was Columbia University's first PhD in anthropology, and the ninth in the US), but these degrees—with their potential to be symbolic capital, seals of quality, and weapons to defend their authority over a field of study—were not yet taken very seriously. In the absence of other means to secure authority in their subject area, concepts (as in the case of the Hippocratics) were their weapon of choice.

Kroeber was outspoken about his goal of defending the exclusive authority of cultural anthropologists over the study of culture, leaving individual development for the aspiring psychologists and nature for physical anthropologists and geneticists. As a result, his case is rather well known. With the exception of Stocking (1968: 259), however, what has often been ignored is that when Kroeber claimed autonomy for culture, he was assuming Weismann's theory of inheritance as proof that the decoupling of culture from nature was possible, and thus that traits caused by culture could be explained independently of nature, and vice versa. I have said more on his case elsewhere (Kronfeldner 2009), but the most important point for our purposes here is that the decoupling he defended rested on a denial of the possibility of the inheritance of acquired characteristics, and was done in the service of disciplinary demarcation and the exclusion of any style of explanation of culture that relied on biology.[14] Like the Hippocratics, he regarded nature as given. However, he gave the attached pragmatic function a negative twist: human nature was taken as a disciplinary primitive; it was again taken for granted, but only in order to have the right to ignore it.[15] Still, 'nature' demarcated kinds of causes (but those to be ignored), united those in the same business of studying a particular kind of cause, and served to demarcate their work from others.

To sum up this case: if we look at the formation of cultural anthropology at the beginning of the twentieth century and compare it to the other cases we have looked at, we see that the enemies have changed, the contrast has changed, but what remained constant was the pragmatic function the term 'nature' (and its contrast) played in demarcating expertise and excluding styles of inquiry.

## 10.10 Moving to the Twenty-First Century

When we move to the twenty-first century, we see the same pattern. Edouard Machery (2008) defends what he calls a 'nomological notion of human nature' (in contrast to essentialist concepts). He discusses why his concept is 'worth fighting for' and writes:

saying that a given property, say a behaviour, such as biparental investment, or a psychological trait, such as outgroup bias, belongs to human nature [...] is also to say that some kinds of explanation for the occurrence of this trait among humans are inappropriate. Particularly, this is to reject any explanation to the effect that its occurrence is exclusively due to enculturation or to social learning.   (Machery 2008: 326)

Machery regards 'explanations [...] exclusively due to enculturation or to social learning' as mere 'proximate explanations', and explanations 'due to nature' as 'ultimate explanations'. According to him, the latter are the appropriate explanations for typical features of human behaviour and cognition.

Irrespective of whether it makes sense, philosophically and scientifically, to divvy up things that way, this fits the historical pattern I describe: Machery uses 'human

---

[14] Interestingly, geneticists joined in with this division of labour—e.g. Thomas H. Morgan and later Dobzhansky—which Richerson (Ch. 8 this volume) describes as agreeing to a 'peace treaty'. This very much fits the picture developed here.
[15] For a philosophical defence of what I have called here 'the right to ignore' certain causal factors—e.g. to ignore human nature or to ignore human culture in one's explanation of behaviour—see Kronfeldner (2017).

nature' for epistemic demarcation, dividing causes into different kinds and defending a specific style of explanation (in his case, a kind of evolutionary psychology) as the appropriate and specialized style of inquiry for the kind of causes—evolutionary causes—that are deemed to be 'natural' rather than 'cultural'.[16]

This kind of appropriation is likely to continue, I reckon, be it in discussion of whether cognitive science has its own way of carving out a concept of human nature (see Heyes, Chapter 4 this volume) or whether anthropology can reclaim human nature for its explanatory goal of studying constrained diversity.[17]

## 10.11 Conclusions

That the concept and the term 'nature', and its contrasts in the context of studying humans, have the pragmatic function of demarcating expertise and excluding styles of inquiry is an important reason why the concept is still with us; the concept can be used to exclude as relevant certain kinds of causal factors in a given context. The contexts I discussed were:

- Hippocratic versus divine healers;
- Enlightenment philosophers defending a 'science of man' versus speculative metaphysics;
- the study of 'hard heredity' versus the Lamarckian approach;
- genetics and physical anthropology versus cultural anthropology;
- evolutionary psychology versus the social sciences.

Further contexts and an open-ended amount of detail could be added to each item on the list, but this kind of historical completeness is not my aim here.[18] To sum up, without reaching any such completeness: even though the methods, the implied contrasts, as well as the content of the term 'nature' (and thus the concept) all varied in the cases mentioned in this chapter, the function of demarcating kinds of relevant causes and styles of inquiry stayed the same.

'Nature' was always what could be taken for granted, established as solid, authoritative, as 'what was presupposed to be there to investigate: its supposed objective reality was what guaranteed the viability of the investigation. Yet what that vaunted objective reality consisted in was contested in every conceivable respect' (Lloyd 1991: 432). I showed that this dictum of Lloyd holds in all the cases surveyed here. In the case of the Hippocratics, some Enlightenment philosophers, and the study of heredity in the

---

[16] Machery has revised his account in a couple of ways (see 2012, and Ch. 1 this volume), but the demarcation remains.

[17] As argued by Fiona Jordan and Heidi Colleran in their paper delivered at the 'Why We Disagree about Human Nature' conference, in Cambridge, 2015, and which gave rise to this volume.

[18] As Raymond Williams (2011: 186) once said with respect to the history of the word 'nature', because of the intricate and many-layered texture of the landscape of contrasts and cognates, 'Any full history of the uses of nature would be a history of a large part of human thought.'

nineteenth century, 'nature' and 'human nature' served to define a style of inquiry (or practice) in a positive sense, as a field-defining explanandum. In the case of Kroeber, it did so in a negative sense, since he regarded human nature as a disciplinary primitive. In all cases, 'nature' demarcated kinds of causes, united those in the same business of studying a particular kind of cause, and served to demarcate their work from others. Demarcation and exclusion is connected with rather non-epistemic pragmatic aims—securing jobs, money, and power—but it is also connected to epistemic issues in two ways: it involves concepts, and it serves to distinguish kinds of causal explanations (e.g. proximate versus ultimate) and the experts devoted to them.

What I have shown also helps us to understand parts of contemporary debates concerning human nature. Machery (2008, 2012, Chapter 1 this volume) defends a descriptive concept of human nature that caters to the needs of certain evolution-minded fields, such as evolutionary psychology, that prioritize evolutionary explanations. Stotz (2010), Griffiths (2011), Lewens (2012), Ramsey (2013, Chapter 2 this volume), and Stotz and Griffiths (Chapter 3 this volume) are sceptical about these needs because it involves the nature/culture divide in a manner that is now contested. They opt for a different concept of human nature, one that is most inclusive and does not divide between kinds of causes, but they ignore that their choice might also serve specific needs. In the case of Stotz and Griffiths, it is the needs of those who care most about explaining development and stressing the importance of development for evolutionary thinking. But that is not the only explanandum in which human nature plays a role. Samuels (2012) defends a concept of human nature that is explanatory in a sense that serves the needs of cognitive scientists, who are not necessarily interested in explaining development. Ramsey (Chapter 2 this volume) is concerned with 'how traits come about', and presents his account as tracking the regularities of the human life form that can also be explained mechanistically. Lewens (2015) is most interested, like Machery, in a human nature concept serving the needs of evolutionary thinking, but he disagrees about which concept of human nature fits these needs, especially if cultural evolution is taken into account.

What is common to most of these authors is that they still look at the issue from a monistic perspective: they look for one concept that will replace the outdated, essentialist concept of human nature. They all want to appropriate the term for their preferred epistemic role, presumably sometimes even to utilize the authority in the term 'nature'. But, I contend, there will be no one concept that fits all the needs of the diverse range of styles of inquiry that employ the term 'human nature'. As part of an essentialist picture (as described in section 10.1, and in more detail in Kronfeldner et al. 2014), all epistemic roles were supposed to be fulfilled by one entity in the world. In the post-essentialist picture, there are at least three scientific 'natures' replacing an essence: a classificatory nature, a descriptive nature, and an explanatory nature. Criteria for membership of a particular form of life, descriptions of forms of life, and causal factors (or mechanisms) of special importance for explaining these forms of life are simply not the same things. These three kinds of natures have an equal right to be regarded as a replacement candidate for the outdated essentialist concept, since they each retain one of the epistemic roles of the essentialist concept. But they can claim only one epistemic role directly; the others are reconstructed indirectly, if at all. Finally, with respect to the explanatory nature, specific disagreements arise because of the complexity of causation (namely, there are always multiple causes involved). Given this complexity, there are different ways to divvy up the totality of developmental and evolutionary causes. Consequently, one can focus on one or another kind of cause as more important, and the focus will depend on the disciplinary affiliation of the experts. It is not an 'anything goes pluralism' that is defended here (as claimed in Stotz and Griffiths, Chapter 3 this volume). Rather, it is a pluralism that applies explanatory parity not only ontologically (at the level of causal factors involved) but also at the level of the multitude of epistemic interests involved in talk about human nature.

This chapter described (rather than evaluated) the authority inherent in the term 'nature' to elucidate one of the reasons we disagree about human nature. I close with a few points connecting to the discussion in this chapter. Some will argue that, as a matter of principle, science should not rely on power and identity politics. Independently of such an in-principle argument, one can also argue that using the term 'nature' for demarcation is no longer necessary, given the disciplinary differentiations firmly established in the architecture of contemporary academia. One might further support such an 'eliminative' stance by stressing that aside from demarcation, the cost of eliminating the term 'human nature' in scientific discourse is rather low. Everything we might want to say about what we above called 'descriptive nature', 'explanatory nature', or 'classificatory nature' could be said without using the term 'nature'. Some, however, might reply that if the term 'human nature' is eliminated from scientific work, others will use it for their goals and do so without meeting the standards of scientific reasoning.[19] Is it not better, then, to appropriate the term for scientific usage? In any case, a detailed discussion of this 'elimination question' must take into account the pragmatic aspects of the term 'nature'. This chapter tried to analyse one such aspect, the authority imbued in the term 'nature'.[20]

# References

Antony, L. M. (1998). '"Human Nature" and Its Role in Feminist Theory.' In J. A. Kourany (ed.), *Philosophy in a Feminist Voice: Critiques and Reconstructions*, 63–91. Princeton, NJ: Princeton University Press.
Antony, L. M. (2000). 'Natures and Norms.' *Ethics* 111: 8–36.
Collingwood, R. G. (1945). *The Idea of Nature*. Oxford: Clarendon Press.
Descola, P. (2005). *Par-delà nature et culture*. Paris: Gallimard.
Foot, P. (2001). *Natural Goodness*. Oxford: Oxford University Press.

---

[19] As argued by Fiona Jordan and Heidi Colleran in their paper delivered at the 'Why We Disagree about Human Nature' conference, in Cambridge, 2015.

[20] For further pragmatic aspects of the elimination question, see Kronfeldner (forthcoming).

Gallie, W. B. (1956). 'Essentially Contested Concepts.' *Proceedings of the Aristotelian Society* 56: 167-98.

Galton, F. (1874). *English Men of Science: Their Nature and Nurture*. London: Macmillan.

Griffiths, P. E. (2011). 'Our Plastic Nature.' In S. B. Gissis and E. Jablonka (eds), *Transformations of Lamarckism: From Subtle Fluids to Molecular Biology*, 319-30. Cambridge, Mass.: MIT Press.

Hacking, I. (1995). 'The Looping Effects of Human Kinds.' In D. Sperber, D. Premack, and A. J. Premack (eds), *Causal Cognition*, 351-83. Oxford: Oxford University Press.

Hager, F. P. (1971-2007). 'Natur: I. Antike.' In J. Ritter, G. Bien, K. Gründer, G. Gabriel, M. Kranz, H. Hühn, and R. Eisler (eds), *Historisches Wörterbuch der Philosophie*. Basel: Schwabe.

Haste, H. (2000). 'Are Women Human?' In N. Roughley (ed.), *Being Human*, 175-96. Berlin: de Gruyter.

Heinimann, F. (1945). *Nomos und Physis: Herkunft und Bedeutung einer Antithese im griechischen Denken des 5. Jahrhunderts*, vol. 1. Basel: F. Reinhardt.

Hippocrates (1849). *The Genuine Works of Hippocrates*. London: Sydenham Society.

Hull, D. L. (1986). 'On Human Nature.' *Proceedings of the Biennial Meeting of the Philosophy of Science Association* 2: 3-13.

Hume, D. (1975). *Enquiries Concerning Human Understanding and Concerning the Principles of Morals*. Oxford: Clarendon Press.

Hume, D. (1978). *A Treatise of Human Nature*. Oxford: Clarendon Press.

Keller, E. F. (2010). *The Mirage of a Space between Nature and Nurture*. Durham, NC: Duke University Press.

Kronfeldner, M. (2009). ' "If There Is Nothing beyond the Organic…" : Heredity and Culture at the Boundaries of Anthropology in the Work of Alfred L. Kroeber.' *Journal of the History of Science, Technology, and Medicine* 17: 107-33.

Kronfeldner, M. (2016). 'The Politics of Human Nature.' In T. Michel Tibayrenc and F. J. Ayala (eds), *On Human Nature: Evolution, Diversity, Psychology, Ethics, Politics, and Religion*, 623-32. Amsterdam: Academic Press.

Kronfeldner, M. (2017). 'The Right to Ignore: An Epistemic Defense of the Nature-Culture Divide.' In R. Joyce (*ed.*), *Routledge Handbook of Evolution and Philosophy*, 210-24. Abingdon: Routledge.

Kronfeldner, M. (forthcoming). *What's Left of Human Nature: A Post-Essentialist, Pluralist, and Interactive Account*. Cambridge, Mass.: MIT Press.

Kronfeldner, M., Roughley, N., and Toepfer, G. (2014). 'Recent Work on Human Nature: Beyond Traditional Essences.' *Philosophy Compass* 9: 642-52.

Lehoux, D. (2012). *What Did the Romans Know? An Inquiry into Science and Worldmaking*. Chicago: Chicago University Press.

Lewens, T. (2012). 'Human Nature: The Very Idea.' *Philosophy and Technology* 25: 459-74.

Lewens, T. (2015). *Cultural Evolution: Conceptual Challenges*. Oxford: Oxford University Press.

Lewis, C. S. (1960). *Studies in Words*. Cambridge: University Press.

Lloyd, G. E. R. (1978). *Hippocratic Writings*. Harmondsworth: Penguin.

Lloyd, G. E. R. (1991). 'The Invention of Nature.' In his *Methods and Problems in Greek Science*, 417-34. Cambridge: Cambridge University Press.

Lloyd, G. E. R. (2003). *In the Grip of Disease: Studies in the Greek Imagination*. Oxford: Oxford University Press.

Lloyd, G. E. R. (2012). *Being, Humanity, and Understanding: Studies in Ancient and Modern Societies*. Oxford: Oxford University Press.

Lloyd, G. E. R. (2015). *Analogical Investigations: Historical and Cross-Cultural Perspectives on Human Reasoning*. Cambridge: Cambridge University Press.

López-Beltrán, C. (1994). 'Forging Heredity: From Metaphor to Cause, a Reification Story.' *Studies in History and Philosophy of Science Part A* 25: 211–35.

Machery, E. (2008). 'A Plea for Human Nature.' *Philosophical Psychology* 21: 321–9.

Machery, E. (2012). 'Reconceptualizing Human Nature: Response to Lewens.' *Philosophy and Technology* 25: 475–8.

Mill, J. S. (1874). 'Nature.' In *Nature, the Utility of Religion, and Theism*, 373–402. London: Longmans, Green, Reader, & Dyer.

Müller-Wille, S., and Rheinberger, H.-J. (2007). *Heredity Produced at the Crossroads of Biology, Politics, and Culture, 1500–1870*. Cambridge, Mass.: MIT Press.

Norton, D. F. (1993). *The Cambridge Companion to Hume*. Cambridge: Cambridge University Press.

Porter, R. S. (1997). *The Greatest Benefit to Mankind: A Medical History of Humanity from Antiquity to the Present*. London: HarperCollins.

Proctor, R. N. (2003). 'Three Roots of Human Recency: Molecular Anthropology, the Refigured Acheulean, and the UNESCO Response to Auschwitz.' *Current Anthropology* 44: 213–39.

Ramsey, G. (2013). 'Human Nature in a Post-Essentialist World.' *Philosophy of Science* 80: 983–93.

Sahlins, M. (2008). *The Western Illusion of Human Nature*. Chicago: Prickly Paradigm Press.

Samuels, R. (2012). 'Science and Human Nature.' *Royal Institute of Philosophy Supplements* 70: 1–28.

Silvers, A. (1998). 'A Fatal Attraction to Normalizing: Treating Disabilities as Deviations from "Species-Typical" Functioning.' In E. Parens (ed.), *Enhancing Human Traits: Ethical and Social Implications*, 95–123. Washington, DC: Georgetown University Press.

Smith, D. L. (2013). 'Indexically Yours: Why Being Human Is More Like Being Here Than It Is Like Being Water.' In R. C. A. Lanjouw (ed.), *The Politics of Species: Reshaping Our Relationships with Other Animals*, 40–52. Cambridge: Cambridge University Press.

Smith, R. (1995). 'The Language of Human Nature.' In C. Fox, R. Porter, and R. Wokler (eds), *Inventing Human Science: Eighteenth-Century Domains*, 88–111. Berkeley: University of California Press.

Smith, R. (1997). *The Norton History of the Human Sciences*. New York: W. W. Norton.

Stocking, G. W. (1968). *Race, Culture, and Evolution: Essays in the History of Anthropology*. New York: Free Press.

Stotz, K. (2010). 'Human Nature and Cognitive–Developmental Niche Construction.' *Phenomenology and the Cognitive Sciences* 9: 483–501.

Thompson, M. (2008). *Life and Action: Elementary Structures of Practice and Practical Thought*. Cambridge, Mass.: Harvard University Press.

Tylor, E. B. (1871). *Primitive Culture: The Origins of Culture*. London: John Murray.

Williams, R. (2011). *Keywords: A Vocabulary of Culture and Society*. London: Routledge.

Zirkle, C. (1946). 'The Early History of the Idea of the Inheritance of Acquired Characters and of Pangenesis.' *Transactions of the American Philosophical Society* 35: 91–151.

# Index

Aardvark 49, 67
Abnormal 93
  Unnatural 71, 93
Acquired characters 59, 100, 197, 200–1
  Acquired variation 146, 148, 152
  Inheritance of 59, 100, 146, 148, 152, 197, 200–1
Adaptation 13, 21, 63, 77–80, 86–7, 96, 99–100, 102, 104–5, 115, 135, 137, 151, 157, 159–60, 164, 172
  Cognitive 104, 159
  Cultural 79, 137, 157
  Genetic 78–80, 86–7
Adaptationist 173
Adaptive 15, 71, 78, 80, 96, 100, 104–5, 115, 131, 134, 158–60
  Adapted 78, 89, 99, 100, 103
  Adaptive plasticity 105
  Adaptiveness 131, 158
Adult 13, 48, 51, 77–9, 85–7, 89, 103–4, 110, 114, 122–3, 129, 136, 139, 160, 164, 170, 178
Affordances, theory of 174–5, 179–81
Aggression 133, 155–6
  *see also* conflict
Alexander, Richard 150, 157–9
Alien 4–5, 8
Allele *see* gene
Altruism 25, 150, 155
  Helping behaviour 150
  Prosociality 20, 120–1, 161
  Reciprocal altruism 150
  *see also* cooperation
American Psychological Society 51
Anatomy 3, 5–7, 10, 23, 146, 190, 195
Anaximander 193
Anaximenes 193
Ancestor 3, 15, 32, 44–5, 82, 115, 131, 155, 165, 177–8
  Ancestral 42, 101, 178, 179, 198
  Ancestry 2–3, 160
  *see also* lineage
Ancient Greece 190–7, 200
  *see also* Hippocratics, the
Animal nature 7, 58, 60–1
Antisocial punishment see *punishment*
Anthropology 2, 12, 29, 51, 59–60, 63, 69, 81, 119, 139, 164, 182, 199–202
Ants 70
Aristotle 28, 88, 188, 193, 197

Artefact 69, 156, 171, 176
Atomism 93
Attention 43, 77, 83, 89, 138–9
  Attention span 43
  Shared (joint) attention 138–9
Autopoiesis 171, 174, 179, 181–2

Babies 70, 115, 123
  *see also* infant *and* foetus
Babylonians 193
Bacon, Francis 195
Bacteria 92, 97, 129
Bamford, Sandra 173
Barnacles 2
Baseball 23, 46, 48–9
Beavers 130
Beckham, David 11
Bees 97
Behaviour information 77
Behavioural
  capacities 14, 22, 32, 81, 98
  development 68, 127–8, 136
  ecology 25, 134, 159
  epigenetics 101
  evolution 150–1
  flexibility 47, 97, 134, 153
  regularities 12, 47, 87–8
  science 20, 29, 33, 133
Behaviourism 86, 151–3
Beliefs 23, 32, 36, 43–4, 78, 82, 88, 136
Bias (evolved) 20, 30, 77, 84–5, 151, 163, 201
  Imitation 77–80, 85, 89
  Learning 163
  Psychological tendency 6, 16, 45, 77, 80, 84, 86, 89, 138, 163
Biochemical 111
  *see also* chemical
Biology, developmental 59, 67–8, 71, 102, 110, 130, 160
Biology, ecological developmental 130
Biology, evolutionary 10, 18–20, 28–9, 32, 36, 72, 108, 114, 133, 135, 140, 145, 148, 151–2, 156–7, 159, 162, 164–5
Biology, evolutionary developmental 29, 130
Biology, systems 66
Bipedalism 24–5, 70, 114–15
Birds 4–5, 68, 79, 128, 130–1
Birdsong 5, 128, 130, 133
Birth (labour) 70, 78–9, 115, 128, 198

Birth, trait present at  78–9, 129, 198
    see also innate and instinct
Blank slate (or tabula rasa)  1, 98, 103–4, 151–2, 161
Blind
    acquisition  83–4, 87
    imitation  81–2
    retention  83–4, 87
    trust  81, 83–4, 86
Blood  19, 48, 129, 135
Boas, Franz  200
Boyle, Robert  93
Brain  70, 101, 121, 128–30, 132, 138, 147, 151, 153, 155, 157, 161–2, 164–5, 174–5, 180
Brain size  70, 155, 161–2, 164

Canalization  21, 63, 66, 70, 113, 128
Canary  130
Campbell, Donald T.  162
Capacities  9–10, 14, 20, 22, 32, 43, 81, 84, 93, 96, 98, 101–2, 112–13, 115–18, 120–1, 131–2, 138, 153–4, 162, 165, 195
    see also disposition and propensity
Cashdan, Elizabeth  46–8
Cats  129
Causal explanatory account  21, 54–5, 88, 127
Causal essentialist theory  87–8
Cavalli-Sforza, Luigi Luca  162–3
Cells  3, 94–6, 100–1, 105, 111, 129–30, 135
Chemero, Anthony  174–5, 179–80
Chemical  49, 53, 96, 101, 105, 199
    see also biochemical
Childhood  51, 80, 86–7, 161
Children  23, 31, 51, 60, 66, 77–9, 81–2, 84–6, 89, 102–4, 112, 114, 116, 119, 121, 132, 136, 140, 153, 161, 170, 174, 177–9, 181–2
    see also offspring
Chimpanzee  6, 53, 60, 112, 115, 118, 121, 186
    see also primates
Chomsky, Noam  161
Christie, Agatha  76
Civilization  1, 8, 147, 152, 200
    see also Western civilization
Classification (of traits as part of human nature)  21, 44, 54–5, 62, 64, 187–9, 197, 203–4
Clothes  5, 129, 136
Cockroach  186
Code see programme
Coevolution see gene-culture coevolution
Cognitive
    adaptation  104, 159
    anthropology  60
    architecture  101–2, 104, 161
    capacity  10, 43, 118
    mechanism  76–7

regularities  12, 47, 87–8
    science  9, 29, 84, 87–8, 97, 151, 174–5, 179, 202–3
Comanche  156
Commonalities (in traits)  7, 9, 13, 62, 67, 127
Communication  68, 77, 80, 89, 115–16, 188
    see also language and speech and song
Conflict  120, 122–3, 156
    Violence  119, 156
    Warfare  120, 150, 155
    see also aggression
Conformist tendency  45, 155, 163, 181
Contingent (response-contingent stimulation)  77–9
Contingent (environmental)  96–7, 103–4, 151, 153, 160
Cooperation  20, 53, 114–16, 118–23, 122–3, 137, 154–5, 180
    see also altruism
Cosmides, Leda  3, 6, 10, 101–2, 134, 148, 159–61, 163
Csibra, G.  13, 77–8, 81–5, 87–8
Culture
    Cultural adaptation  79, 137, 157, 163
    Cultural ancestry  160
    Cultural determinism  201
    Cultural evolution  12–13, 15, 33, 35–6, 77, 81, 83–4, 88, 103, 137–8, 146, 149–50, 152–8, 160–5, 203
    Cultural forces  15, 100
    Cultural influence  6, 14, 30–1, 34, 43, 54, 89, 92, 120, 128, 156, 164, 200
    Cultural inheritance  12–13, 77, 80–5, 88–9, 135, 138, 153, 163
    Cultural practice  70, 81, 122, 138, 140, 170, 172, 176–7, 188
    Cultural processes  7, 11–12, 15, 30, 35–6, 88, 136, 138, 146, 149–50, 152, 154–7, 161, 163–5, 171
    Cultural transmission  20, 33–4, 83–4, 103, 120, 122, 135, 138, 155, 162, 171
    Cultural variants  33, 155, 163
    Culturgens  162
    Enculturation  30, 35, 42–3, 201
    Nomos  196–7
    Custom  147, 178, 197
        Convention  7, 82
        Habit  154, 195
        Tradition  12, 147, 163
Cuvier, Georges  98
Creativity  98, 104
    see also imagination

Dairy  139, 164
    see also farming
Darwin, Charles  2, 6, 15, 63, 98–9, 103, 108, 146–50, 152–6, 162, 198–9

Darwinism 58, 100–1, 145, 150, 153, 160, 189
  see also Modern Synthesis
Dawkins, Richard 162
Dehumanization 187
Demarcation (of expertise) 187, 191–7, 199–204
Democracy 53–4
Democritus 93
Dennett, Daniel 78, 97
Descendant 109, 131, 135–6
  Descent with modification 160
Determinism 60–1, 76, 98, 117, 135, 155, 200
  Cultural 177, 200
  Genetic 60–1, 117, 135
Developmental
  constraints 32, 48, 65, 67, 70–1
  cycles 95
  environment 14, 59, 70–2, 112, 131, 136
  processes 12, 13, 15, 30, 59, 66, 68–9, 94–5, 105, 160
  trajectory (or pathway) 11, 13, 65–7, 71, 95, 102, 104–5, 110, 160
Developmental system theory 59, 64–71, 110, 130, 175, 179
Devitt, Michael 109–11
Dichotomies (or distinctions)
  Biology/culture 58, 173, 174–6, 180, 182
  Genes/environment 129
  Germ/soma 199
  Innate/acquired 30, 58–9, 116, 129
  Mind/body 174
  Natural/supranatural 191–7
  Nature/culture (or biology/culture) 15, 58, 170, 173–6, 180, 182, 187, 191, 197–8, 200, 203
  Nature/nurture 58, 128, 132, 149, 172, 198–9
  Proximate/ultimate 33, 110, 136–7, 146, 165, 201, 203
Disease 11, 60, 112, 153, 192–4, 196–9
  see also health
Disposition 7, 20, 22–3, 28, 37, 41, 44, 86, 103, 199
  Pre-disposition 20, 86
  see also capacity and propensity
Divine 192–4, 202
  see also religion
Division of labour 116, 120, 133
  see also social, arrangement
DNA 19, 60, 135, 171
Dobzhansky, Theodosius 148–52, 157, 165, 201
Dog 2, 6, 121
Dominance (behavioural) 23, 133
Dove 128
Drake, Frank 4
Drosophila 133
Ducks 118, 121, 128

Economics 9, 12, 20, 51, 53–4, 118, 120, 179
  see also game theory
Education see learning and teaching
Egypt 193
Elephant 113
Eliminative account of human nature 2, 58, 76, 204
Emotion 22, 44, 80, 113–15, 121, 153, 161
  Anger 22, 41
  Courage 156
  Disgust 22, 34
  Docility 138
  Excitement 45
  Fear 6–7, 24, 101, 161
  Guilt 113
  Happiness 22, 45
  Jealousy 22
  Pride 113
  Shame 22, 24, 44, 113
Empathy 13, 121, 178
Enlightenment 191, 194–6, 202–3
Epigenetics 12, 62, 66–7, 69–70, 88, 100–1, 103, 135, 162–3, 165
Epistemological resources (in development) 69, 103, 132
Epistemology
  Epistemic authority 187, 191–3, 195–7, 199–200, 202, 204–5
  Epistemic role of the concept of human nature 16, 62, 71, 186, 188–9, 194–5, 203–4
  Epistemic toolkit 10, 71, 193–4, 196
Essence 19, 25, 41–2, 47, 54–5, 58, 67, 92, 109, 111, 133, 186, 188, 191, 203
Evolutionary design 59–64, 137, 159
  see also teleology
Evolutionary origins 42, 84–5, 87–8, 127, 151, 172
Evolutionary psychology 3–4, 25–6, 87, 95, 100–1, 104, 109, 117, 134–7, 139, 145, 148, 192, 202–3
Evolutionary social science 119, 162
Exaptation 137
Experiment 9, 60, 65, 71, 97, 120, 128–9, 138, 161, 176, 180, 195, 198–9
Expertise (disciplinary) 191–2, 194, 197–8, 201–4
Experts (as models in development) 80–4, 115
Explanation
  Causal 21, 54–5, 60–1, 88, 127, 194, 203
  Cultural 36, 92, 100, 116
  Developmental 110
  Evolutionary 23, 32–3, 36–7, 69, 110, 116, 165, 203
  Mechanistic 95, 111, 203
  Proximate 201, 203
  Structural 110–11

Explanation (*cont.*)
  Ultimate 32, 44, 54, 116, 136–7, 146, 163, 165, 201, 203
  Extended evolutionary synthesis 131, 146, 165
Eye contact 77–80, 85
  *see also* gaze

Farming 131, 139
  *see also* dairy
Feedback loop *see* reciprocal causation
Feldman, Marcus W. 103, 162–3
Fidelity (in inheritance) 77, 81, 83–6, 122
  One-shot 83–4
  Recurrent 83–6
Fiji 170, 176, 178
Fire 23, 113–14, 129, 131
Fisher, R. A. 159
Fitness 44, 69, 79, 100, 150, 154–5, 162–3
Flexibility *see* plasticity
Foetus 70, 93
  Embryo 70, 131
  *see also* babies
Folk tales 35
Food 23, 97, 103, 131–2, 138, 163
Foraging (or gathering) 133, 156, 164

Gait 3, 114–15
Galapagos finches 160
Galileo 195
Galton, Francis 28, 198–9
Game (animal) 131
Game (playing) 32, 48, 129
Game theory 53, 120, 151
  *see also* economics
'Gangnam Style' (Psy) 33, 36, 45
Garland, Hamlin 46
Gautier, David 122
Gaze 77–80, 85, 89
Geertz, Clifford 1
Gemmule 199
Gender 27, 51
Gene
  Allele 100, 112, 139, 163
  Genome 49, 59, 67, 70, 100–1, 111–12, 131, 157
  Genotype 47, 71, 131, 150, 162–3
  Gene-centric 14, 134, 172
Gene-culture coevolution 13–14, 35, 116, 137–9, 146–7, 154–5, 157–8, 161–5
Genealogy 2–3, 19, 67, 108, 110, 113, 187–9
Genetic drift 10, 32, 99–100, 137
Genetic mutation 10, 100, 102, 112, 137
Genetic programme *see* programme
Genetic variation 11, 15, 85, 111, 134, 137–8, 151, 155
Genetics 10, 18–20, 29, 36, 69, 137–8, 149–50, 162–3, 173, 199, 202

Gergely, G. 13, 77–8, 81–5, 87–8
Germline 101, 135
Gesture 82, 177
Ghiselin, Michael 3, 5, 145
Godfrey-Smith, Peter 5, 108, 113, 123
Gray, Henry 6
*Gray's Anatomy* 6–7, 10
Greek antiquity 190–4, 196–7, 200
Groups
  Biological group 2, 27, 61, 69, 188
  Group living 116, 157
  Social groups (or human groups) 7, 15, 22–3, 26–7, 37, 42, 60–1, 84, 86, 103, 116, 120, 122, 133–4, 150, 155–6, 158, 164, 187–9, 191
  In-group 61, 122, 187
  Out-group 30, 34, 187, 201

Habit 147, 154, 195
Haldane, J. S. 105
Hamilton, W. D. 150, 157–8, 163
Health 66, 130, 172
  *see also* disease
Heraclitus 93
Hereditary 153, 198–9
Heredity 66, 131, 153, 191, 198–9, 202
Heritable 30, 69, 153, 163
Heuristic 86
Hippocratics, The 192–4, 196–7, 200–2
  *see also* Ancient Greece
Hobbes, Thomas 196
Holocene 119, 160
Homeostasis 105
Hominin 16, 82, 85–6, 114–16, 131
*Homo sapiens* 2–4, 9, 12, 15–16, 19, 24, 41, 113, 147, 155, 157, 164, 188–9
Hull, David 1, 3, 41, 47, 55, 62, 108–9, 111–13, 121, 145, 156–7, 186, 188
Human condition 66, 127, 139–40, 157, 159–60
Human diversity (or variation) 9, 12, 26, 28–9, 52, 59, 64, 67, 70–1, 93, 109, 119, 134, 139, 140, 202
Human nature
  Anti-essentialism about 2–4, 47, 109, 111, 113, 188–9
  Causal essentialist account 47, 87–8
  Classificatory concept of 62, 64, 187–9, 197, 203–4
  Descriptive concept of 20–1, 24–6, 53, 59, 62, 65–6, 68, 114, 123, 187–90, 195, 197, 203–4
  Elimination of 2, 58, 76, 204
  Essentialist (traditional) concept of 2–4, 7, 18–19, 36, 41–2, 54–6, 58, 60–1, 108–11, 123, 132, 145, 151, 156–7, 186–90, 201, 203–4

Evolutionary causal essentialist account 12, 77, 88–9
Explanatory concept of 12, 20–1, 29, 36, 54–5, 59, 62, 65, 88, 114, 116, 120, 187–90, 195, 197, 202–4
Field-guide concept of 4–6, 113, 123
Folk-biological concept of 58–64, 66, 71, 187
Humean concept of 15
Life-history trait cluster (LTC) account 11, 41–2, 46–56, 59, 64–7, 69, 71, 189
Limitations imposed by 14, 21, 50, 71, 117, 119, 156, 159
Nomological account 9, 18–26, 28–36, 43, 46, 50, 52–6, 87–9, 117, 201
Normative implications of 9–10, 50, 61–2, 65–6, 71, 88, 112, 114, 117, 121, 160, 182, 186, 188–90
Pluralism about 11, 71–2, 176, 188–9, 204
Pragmatic function of 16, 34, 43, 71–2, 190–6, 200–4
Scepticism about 2, 9, 13–15, 41, 47, 117
Three-factor model of 58, 60–1
Trait bin account 11–12, 41–2, 44–7, 49–52, 54–6
Human sciences 2, 12, 41, 50–2, 55–6, 59, 63–4, 71, 170, 172, 174
Hume, David 7–9, 88, 195
Hunting 133, 154–6
Hunter-gatherers 151–2, 156, 164

Imagination (human trait) 98, 113
*see also* creativity
Imitation 6, 20, 77–86, 89, 132, 138–9, 147, 153, 164
Immigrants 200
*see also* migration
Immunity 96, 129, 136
Imprinting 79, 130
Incest avoidance 24, 133, 137
Infant 13, 70, 76–85, 89, 115, 119, 122–3, 132, 136, 138, 153, 161, 165, 181
*see also* baby
Infanticide 123
Inferential process 77–8, 195
Information
Behaviour information 77, 81, 89, 118
Cultural information 115, 132
Evolutionary information 32–7, 45
Genetic information 129
Ontogenetic information 32–3
Inheritance
Of acquired characteristics (Lamarckian) 100, 146, 148, 152–4, 160, 197, 199–202
Biological inheritance 98, 153
Cultural inheritance 12–13, 77, 80–5, 88–9, 135–6, 146, 153, 163

Ecological inheritance 69, 103, 131, 135
Epigenetic inheritance 12, 66, 69–70, 88, 100–1, 103, 135, 165
Genetic inheritance 11, 13, 59, 66–7, 70, 77, 81, 103, 136, 149, 153, 158
Hard inheritance 199, 202
Non-genetic (extended) inheritance 14, 59, 69, 136, 146
Innate 5, 20, 30, 34, 58–9, 64–6, 101, 103, 115–16, 119, 128–9, 132, 146, 151–3, 156, 161, 170
*see also* birth, present at *and* instinct
Innovation 100, 115–16, 148, 159, 164, 175
Biological 175
Cultural 148, 159, 164
*see also* novelty
Instinct 101, 128–9, 151
*see also* innate
Institutions (as cultural resource/trait) 63, 81, 98, 116, 122
Intelligence 3–4, 97–8, 102–3, 147, 153–4, 157, 160, 165
Intentionality 84–5, 121, 179, 189
Intentional stance 78, 85
IQ 150

Jablonka, Eva 100
Jepsen, Carly Rae 33, 45
Juveniles 61, 103, 122
*see also* children *and* offspring *and* youth

Kail, Peter 7
Kangaroo 60
Keller, Evelyn Fox 198
Kimura, Motoo 100
Kin 25, 116, 133, 157–8, 161, 178–9
Kinds
Etiological 54
Human 63, 188–9, 201
Natural 108
Scientific 49, 108, 110
Kitcher, Philip 122
Knowledge production 190
Kroeber, Alfred L. 200, 203

La Mettrie, Julien Offray 196
Lactose *see* dairy
Lamarckian inheritance *see* inheritance, of acquired characters
Language 3, 13, 27, 43, 69, 78, 81–2, 84, 93, 112–13, 115–16, 121–2, 132, 137–8, 153, 156, 161, 174, 177–9, 189
*see also* communication *and* speech
Law (legal) 7, 137, 147, 197
Law (of nature) 31, 49, 111, 151, 193, 195, 198
Le Guin, Ursula 118
Learnability problem 82

Learning  5, 10–13, 15, 20, 30–6, 42–3, 45–6, 53–4, 65, 70, 77–89, 97–8, 101–4, 109, 115–16, 120, 128–30, 132, 135–6, 138, 146, 151, 153, 159–60, 163, 165, 170, 174, 201
  see also mechanisms, for learning and natural pedagogy and teaching
Lehrman, Daniel  128–9
Lewis, David  122
Life-history strategy  64, 69
Lineage  13, 16, 19, 42, 65, 66, 95, 99, 100, 102, 103, 105, 109, 110, 111, 113, 114, 115, 116
  see also ancestry
Linguistics  29, 161, 176
Lloyd, Geoffrey  173, 176, 188, 190–3, 202
Locke, John  88, 93
López Beltrán, Carlos  198
Lumsden, Charles J.  154, 157, 162–4

Machines (humans as)  98, 196
Marriage  161
  see also sex (behaviour)
Maternal care  42, 48, 101, 116, 129, 131, 161, 181
Mates see sex (behaviour)
Maynard Smith, John  150–1, 157, 163
Mayr, Ernst  33, 110, 136–7, 146, 165
Meaney, Michael  101
Mechanism (cognitive)  7, 12–13, 47, 76–80, 82, 87–8, 134, 151
Mechanism (philosophical view)  95–6, 111, 203
Meditation  51–2, 55, 130
Meme  162
Memory  43, 101, 113, 115, 121, 130
Menarche  48
Menopause  26, 43, 48, 114
Merleau-Ponty, Maurice  170, 179
Metabolism  70, 95–6, 99, 102, 105, 115, 129
Mice  6
Microbes  96, 129
Microhistorical ontogenetic process  15, 170, 172, 175, 179–80
Migration  10, 163–4
  see also immigrants
Mind-reading see theory of mind
Mitchell, Sandra  72
Models (formal)  20, 28–9, 33, 118, 120, 134, 137–8, 150, 154, 162–3
Modern Synthesis  14–15, 100, 146, 148–9, 152–4, 156–63, 165
  see also Darwinism
Modularity  95, 100, 115, 160
Monism  203
Montague, Ashley  150–1
Moral psychology  109

Morality
  Capacity for  13, 109, 113, 147
  Moral behaviour  20, 53, 120, 122, 133, 161, 188–9
  Moral progress  147, 155
  Norms  9, 20, 32, 109, 113, 117, 121–2
Morphology  26, 87, 92, 96–7, 109–10, 116, 118–19, 133, 138
  see also polymorph
Mother Nature  86, 89
Music  9–10, 31–3, 45, 116

NASA  4, 8
Natural history  104
Natural philosophy  193, 195
Natural pedagogy  13, 76–89
  see also teaching and learning
Natural-state models  28–9
Naturalism  60, 123, 151, 191–5
Neanderthals  132, 188
Needham, Joseph  93
Neural system  6, 45, 128, 130, 135, 153, 156, 175, 180
Neurophysiology  45, 156, 180
Neuroscience  31
Neutral theory  99–100, 102
Nest
  Ant  70
  Bee  97
  Bird  128, 131
  Wasp  131
Newton, Isaac  195
Niche  14–15, 19, 32, 59–60, 66–71, 97–8, 102–4, 130–2, 135, 137, 140, 154–5, 165
Niche construction  14, 32, 59, 69–71, 97, 103, 130–1, 135, 137, 140, 154–5, 165
Norm of reaction  47, 71
Novelty  100, 158
  see also innovation
Nurture  11, 58, 128–30, 132, 139, 149, 172, 175, 198–200

Odling-Smee, John  103, 131
Offspring  69, 98–9, 101, 103, 131–2, 135–6, 140, 158, 198
  see also children and juvenile and youth
Old age  23, 93, 178
Ontology
  Darwinian  189
  Process  13, 15, 59–60, 66, 68–9, 93–6, 99–105, 130, 135, 140, 175, 187
  Substance  93–4, 103–4, 187
Oyama, Susan  175, 179

Parents  22–3, 60, 69–70, 88, 103–4, 116, 131–2, 135, 153, 157, 171, 198, 201
  Biparentality  22–3, 202

Child rearing 70, 102, 132
Parental care 70, 132, 158, 202
Parental effects 69, 132, 136
Parmenides 93
Pathology *see* disease
Peace 114, 122, 156
Pedagogy *see* natural pedagogy
Perception 3, 6, 7, 9, 32, 60, 85, 112, 174–5, 180, 182
Phenotype 12, 45, 54, 60–3, 69, 71, 100–2, 108–10, 112, 121, 127, 131–2, 135–6, 147, 151, 155, 158
 Developmental phenotype 131
 Phenotypic flexibility (plasticity) 63, 69, 147, 151, 155
Phylogeny 21, 32, 55, 78, 80, 156, 160
 Cultural phylogeny 161
Physiology 3, 5–6, 10, 12, 23, 29, 59, 63, 87, 92–3, 109, 118, 146, 161, 171, 190
Piaget, Jean 153, 170, 179, 181
Pinker, Steven 1, 3, 148, 161
Pioneer 10 4–5, 8
Plasticity 12–13, 47, 59, 63–4, 66–9, 71, 96–8, 100, 102, 104–5, 109, 113, 115, 128, 130, 134–5, 146, 150, 153–4
Playing *see* game (playing)
Pleistocene 117, 119, 155–6, 159–61, 164
Polymorph 40, 59, 63
 *see also* morphology *and* sexual dimorphism
Population-thinking 28–9, 37, 162–3
Primate 3, 32, 70, 81, 112, 114–16, 118–20, 133, 154–5, 161, 164–5
 *see also* chimpanzee
Programme
 Code 67, 130, 153, 162
 Developmental programme 36, 160
 Genetic instructions 135
 Genetic programme 98, 128, 134–5, 153, 160
Propensity, evolved (or tendency) 6, 14, 16, 45, 77, 80, 84, 86, 89, 133, 138, 151, 158–9, 163
 *see also* capacity *and* disposition
Property cluster account *see* human nature, life-history trait cluster account
Prosociality *see* altruism *and* cooperation
Proteins 68, 101, 111, 130
Proust, Marcel 170
Proximate
 Cause 33, 136–7, 146, 165
 Explanation 201, 203
 *see also* ultimate
Psy 33, 45
Punishment 53, 120, 155, 163
 *see also* reward
Psychic unity 148
 *see also* universal traits

Quetelet, Adolphe 28

Rationality 78, 80, 120, 155, 188–9
Rats 68, 101
Reasoning 6–7, 9, 13, 120, 195
Reciprocal altruism *see* altruism
Reciprocal causation (feedback loops) 14, 86, 115, 120, 130, 135, 149, 158
Reductionism 72, 151
Reiss, John 98–9
Religion 50, 113, 147, 155, 179
 *see also* divine
Replication 158
Reproduction 97, 99, 102, 105, 111, 113–15, 149, 158, 191, 197–8
Reward 85–6, 89, 156, 164
 *see also* punishment
Risk aversion 23
Rivers, W. H. R. 9
RNA 94, 131, 135
Rose, Nikolas 172–4
Rousseau, Jean-Jacques 196

Sagan, Carl 4, 8
Sahlins, Marshall 1, 187
Salzman Sagan, Linda 4, 8
Samuels, Richard 12, 46–7, 87, 203
Scaffolding 103, 120, 135
Schneirla, Theodore 129
School 11, 31, 44, 98, 103
 *see also* learning *and* teaching
Scientific progress 36, 67, 160
Seabright, Paul 122
Selection
 Group 150, 155, 157–8
 Kin 157
 Natural 2–3, 10, 15, 32, 62, 70, 77, 79, 80, 98–100, 102, 130, 134–5, 137–8, 146–7, 150–2, 155–7, 159, 162–3, 165, 175
 Social 152, 154–5, 162–4
Sex (behaviour) 117–18, 122, 128, 149, 155
 Mates 19–20, 23, 155
 Courtship 128, 133
 Pair bond 153, 197
Sex (class) 8, 26–7, 60, 117–18, 133, 160
Sexual dimorphism 26, 160
 *see also* polymorph
Singer, Peter 121–2
Skills 35, 77, 81, 104, 114, 129, 136, 165, 177
Smith, Roger 194–5
Social
 arrangement (or organization or structure) 14, 69, 114, 116–20, 133, 160
 behaviour 20, 68, 111, 120, 149, 152–3, 156–7, 159, 161
 bonding 79–80, 89, 133
 convention 7, 82

Social (*cont.*)
  environment 14, 31, 44–5, 98, 115, 117, 119–20, 123, 154
  life 16, 20, 53, 111, 115, 117, 119, 121
  motivation 7, 84–7, 89, 115, 117–18, 138
  relations 14, 19, 63, 117, 171, 174, 176–82, 189
Social construction (of human nature) 14–15, 140
Social science 1, 52, 58, 104, 109, 119, 149–53, 157–8, 161–3, 165, 175, 203
Society 9, 14, 53–4, 119–20, 122–3, 134, 153, 155–6, 158, 161
Sociobiology 117, 133–4, 149, 151–2
Sociology 31–2, 51, 59, 63, 104, 190
Speech 22, 33, 45, 77–80, 82–3, 89, 93, 112, 132, 136, 138, 171, 177, 180–1
  *see also* communication *and* language
Spiders 130
Standard social science model 104
Status, social 53, 178
Stirps 199
Strangers 53, 117, 119, 122, 133, 161, 180
Strathern, Marilyn 173, 176, 182
Stress (or anxiety) 52, 55, 113, 130, 181
Superorganic 15, 149–50
Supranatural 191–7
Symbionts 96, 129, 131, 135

Tabula rasa *see* blank slate
Teaching 7, 13, 21, 76–81, 84, 89, 102, 116, 132, 138–9, 147
  *see also* learning *and* school
Technology 3, 82, 103, 122, 129, 138, 160
  *see also* tools
Teleological theory 78, 81–2, 84
Teleology 58–64, 84, 197
  Telos 197
  *see also* evolutionary design
Tendency *see* propensity, evolved
Termites 97, 130–1
Theory of mind 78, 84, 132–3
Thompson, Evan 175, 179–80
Tomasello, Michael 84, 174, 179
Tooby, John 3, 6, 10, 101–2, 134, 148, 159–61, 163
Tools 69, 81–4, 132, 138, 154, 157, 164
  *see also* technology
Trade
  Barter 136

  Market 118, 120
  Money 136
Transmission *see* inheritance
Trivers, Robert 150–1, 157–8, 163
Tylor, Edward Burnett 200

Ultimate
  cause 33, 136–7, 146, 154
  explanation 32, 44, 54, 116, 136–7, 146, 163, 165, 201, 203
  *see also* proximate
Underdetermination 189
UNESCO 150
United States 200
Universal traits 3, 9–10, 12, 14, 42, 47, 63–4, 70–7, 76, 101–4, 109, 112–13, 121, 127, 129, 133–7, 139, 151, 153, 159, 161, 182
  *see also* psychic unity
Universality proposal 18, 27–8, 37, 87
Unnatural *see* abnormal
Utopia 117

Values 88, 156
  *see also* moral, norms
Variant
  Cultural 33, 137, 155, 163
  Developmental 28, 131
  Epigenetic 135
  Genetic 77, 163
Vision *see* perception
von Bertalanffy, Ludwig 93

Waddington, C. H. 93, 105
Wallace, Alfred Russel 99, 199
Wasp 97, 131
Weismann, August 199–200
West-Eberhard, Mary Jane 97
Western civilization 1, 9, 173, 191
  *see also* civilization
Whitehead, Alfred North 93–4, 105
Williams, George 150–1
Williams–Beuren syndrome 28
Wilson, E. O. 133–4, 149, 151–9, 161–5
Wittgensteinian criteria 96

Xenophanes 193

Yoga 130
Youth 99, 147
  *see also* children *and* juvenile *and* offspring

The manufacturer's authorised representative in the EU for product safety is
Oxford University Press España S.A. of el Parque Empresarial San Fernando de
Henares, Avenida de Castilla, 2 – 28830 Madrid (www.oup.es/en or product.
safety@oup.com). OUP España S.A. also acts as importer into Spain of products
made by the manufacturer.

www.ingramcontent.com/pod-product-compliance
Ingram Content Group UK Ltd.
Pitfield, Milton Keynes, MK11 3LW, UK
UKHW022136220326
469240UK00007B/69